CW00375622

Le
from
Gretley

Mindful
leadership
and the law

Disclaimer

No person should rely on the contents of this publication without first obtaining advice from a qualified professional person. This publication is sold on the terms and understanding that (1) the authors, consultants and editors are not responsible for the results of any actions taken on the basis of information in this publication, nor for any error in or omission from this publication; and (2) the publisher is not engaged in rendering legal, accounting, professional or other advice or services. The publisher, and the authors, consultants and editors, expressly disclaim all and any liability and responsibility to any person, whether a purchaser or reader of this publication or not, in respect of anything, and of the consequences of anything, done or omitted to be done by any such person in reliance, whether wholly or partially, upon the whole or any part of the contents of this publication. Without limiting the generality of the above, no author, consultant or editor shall have any responsibility for any act or omission of any other author, consultant or editor.

Lessons from Gretley

Mindful leadership and the law

Andrew Hopkins

CCH

a Wolters Kluwer business

CCH AUSTRALIA LIMITED

GPO Box 4072, Sydney, NSW 2001

Head Office North Ryde
Phone: (02) 9857 1300 Fax: (02) 9857 1600

Customer Support
Phone: 1 300 300 224 Fax: 1 300 306 224
www.cch.com.au

Book Code: 34094A

ABOUT CCH AUSTRALIA LIMITED

CCH Australia is a leading provider of accurate, authoritative and timely information services for professionals. Our position as the "professional's first choice" is built on the delivery of expert information that is relevant, comprehensive and easy to use.

We are a member of the Wolters Kluwer group, a leading global information services provider with a presence in more than 25 countries in Europe, North America and Asia Pacific.

CCH — *The Professional's First Choice.*

Enquiries are welcome on **1300 300 224**

National Library of Australia Cataloguing-in-Publication Data

> Hopkins, Andrew, 1945–.
> Lessons from Gretley : mindful leadership and the law.
>
> Bibliography.
> Includes index.
> ISBN 978 1 921223 31 0 (pbk).
>
> 1. Gretley Mine Disaster, Newcastle, NSW, 1996. 2. Coal mines and mining – Law and legislation – New South Wales – Cases. 3. Mine safety – Law and legislation – New South Wales – Cases. 4. Employers' liability – New South Wales – Cases. 5. Judgments – New South Wales. 6. Coal mine accidents – New South Wales. I. CCH Australia Limited. II. Title.
>
> 343.944077520264

© **2007 CCH Australia Limited**

All rights reserved. No part of this work covered by copyright may be reproduced or copied in any form or by any means (graphic, electronic or mechanical, including photocopying, recording, recording taping, or information retrieval systems) without the written permission of the publisher.

Printed in Australia by McPherson's Printing Group

Preface

The Gretley mine disaster took the lives of four men and devastated their families and friends. The people subsequently convicted were also devastated. This work seeks to salvage what it can from the devastation.

The book is written with multiple audiences in mind. In part, it is addressed to those concerned with regulatory policy — what kinds of legal/regulatory strategies can be used to focus the minds of leaders on safety? In part, it is addressed to organisational leaders — what can they do to maximise the chances of safe operation? In the language of the book, what does it mean to manage mindfully? Beyond these specific groups, this book is directed at the broader readership of people interested in the way in which organisations can be made to function more safely.

At the time of writing, the NSW Government is wrestling with claims and counter-claims about the need to change the state's OHS Act. I hope that *Lessons from Gretley* will contribute to that debate.

The manuscript, in whole or in part, has been read by John Braithwaite, Neil Foster, Neil Gunningham, Tony Honoré, Anthony Hopkins, Tamar Hopkins, Heather McGregor, Sally Traill and Sonya Welykyj. I am grateful to all of these people for their comments and corrections. I am especially grateful to Neil Foster for the interest he has shown in this project. Thanks to Sonya Welykyj for her meticulous research assistance. My wife, Heather McGregor, has yet again graciously tolerated my absent-mindedness. The project was financially supported by the Australian Safety and Compensation Council.

Andrew Hopkins

Canberra

(Andrew.Hopkins@anu.edu.au)

February 2007

CCH Acknowledgments

CCH Australia Ltd wishes to thank the following team members who contributed to this publication:

Product Director: Laini Bennett

Editor: Deborah Powell

Product Analyst: Karyn Ashlin

Production: Fergus Ong, Ravinderen Kandiappan and Safiyyah Ahmad Sabri

Graphics: Scott Collins

Indexing: Mark Southwell

Marketing: Suzanne Jammal

Cover: Trina Hayes, Feathered Edge Design

CONTENTS

Chapter 1

Introduction

Disaster struck at Gretley coal mine near Newcastle in November 1996. Miners inadvertently broke through into the flooded tunnels of an abandoned mine, and a wall of water rushed in with enormous force. A mining machine, weighing nearly 50 tonnes, was pushed 18 metres along the tunnel, until it jammed sideways. Four men were swept to their deaths.

The Gretley disaster provoked an unprecedented legal response. Three mine officials — the current mine manager, a former manager at the mine, and the mine surveyor — were convicted and heavily fined. Never before in Australia had managers been held accountable in the way that the Gretley managers were, and the convictions sent shock waves through Australian industry. Did this mean that all Australian managers were now at risk of prosecution when workers were killed?[1]

The convictions triggered a strenuous campaign by employer groups against the legislation[2] and, at the time of writing, the New South Wales Government had still not decided how to respond. The Gretley prosecution may therefore turn out to have been an exceptional event, brought about by an exceptional set of circumstances. On the other hand, there is widespread support for holding directors and managers more accountable for safety, and the Gretley prosecution may herald a new era in this respect.

Whether the prosecution was an anomaly, as employers hope, or a turning point, as unions hope, it is certainly worthy of attention. It lays bare the workings of the legal system in response to workplace fatalities and it raises in an acute way a number of issues about the legal system's role in encouraging workplace safety. It forces us to think about the meaning of justice and the purposes of the criminal justice system. It poses questions about the culpability of individual managers when things go wrong. It ranks with the prosecution of Esso (following its gas plant explosion at Longford) as one of the landmark cases in Australian OHS law (Hopkins, 2002). This introduction aims to identify, in a little more detail, the themes that this book will pursue.

Leaders

At the most general level, this book is about the role that the law plays in ensuring that leaders attend to safety. Is it reasonable to hold leaders accountable

---

1   The convictions were varied on appeal, but not in a way that provides much comfort to managers. *Newcastle Wallsend Coal Company Pty Ltd & Ors v Inspector McMartin* [2006] NSWIRComm 339 (5 December 2006). See ch 2, fn 38.

2   *Occupational Health and Safety Act 2000* (NSW).

for the safety of workers? What is the best way for the law to achieve this end? These are the questions to which I hope to suggest answers.

Safety is primarily a leadership responsibility. It is true that workers have a role in ensuring their own safety and the safety of others, but it is leaders who establish the culture of an organisation (Schein, 1992), allocate resources and establish priorities. These things are crucial to workplace safety and they are peculiarly within the control of leaders. That is why it is appropriate to focus on the role of the law in encouraging leaders to attend to safety.

The word "leader", used in the preceding paragraphs, is deliberately vague. It refers most obviously to company directors and senior executive officers, but it also refers to leaders of organisational units that are themselves part of a larger corporate structure. It may well be appropriate to regard the manager of a large industrial site as a leader, even though he or she is answerable to higher level corporate leaders. Whether it is reasonable to describe people as leaders will depend entirely on the organisational context. By using the word "leader", I seek to draw attention to the role that such people play, wherever they may be in an organisational hierarchy.

An author's dilemma

As an author, I face a dilemma. Should I take a general question as my organising principle, or should I organise the material around a particular case study? There is a logical appeal and a certain simplicity about pursuing a general question and drawing on whatever material may be useful in answering it. On the other hand, there is a fascination about a case study. It brings things to life and gives them an immediacy, an urgency, that more abstract discussions almost inevitably lack. However, there is one aspect of the case study approach that must be noted: a case study takes on a life of its own. It raises issues beyond the initial motivating question and it compels attention to these emergent issues, even though they are in some respects tangential to the original concern. Every time I write a book, I face this dilemma of how to organise my material, and every time I seem to come down on the side of the case study.

This book, then, is a case study.[3] The question that interested me as I sat down to write was the role that the legal system could play in focusing the attention of leaders on workplace safety. The Gretley case provides a context for addressing this question, but it also provokes various questions that are specific to the case: why did this prosecution take place at all; just how culpable *were* the convicted individuals; and what can we learn about mindful management from the failure of Gretley's risk management strategies? Given that the Gretley case is the organising principle for this work, these wider questions are addressed here.

---

3   For a detailed discussion of the limitations and strengths of case studies, see Flyvberg (2001, ch 6).

The demand for punishment

One of the most striking things about the Gretley prosecution was the fact that it occurred at all. No Australian mine disaster had previously culminated in the prosecution of a mining company, let alone the prosecution of individual managers. Yet, on this occasion, relatives of the victims demanded incessantly that managers be held accountable for the deaths. It was for them a question of justice. In the end, the government agreed that justice required the mine and its managers be prosecuted. The government's decision to prosecute was in no sense a foregone conclusion. It was the outcome of a distinctive series of political circumstances to be described later. The somewhat unusual nature of these circumstances is one of the reasons for doubting whether the pattern of the Gretley prosecution will occur again in the near future.

Justice, of course, is one of the purposes of the criminal justice system, but justice is not a self-evident concept. Why does justice require punishment? And what precisely was it that the relatives were wanting when they demanded justice? When probing this question in the Gretley case, it turns out that what the relatives wanted was some form of closure that would provide them with whatever peace of mind could be recovered from the situation. Justice for them was a means of achieving this peace of mind. Importantly, for the relatives, the question of whether or not the managers were actually culpable was hardly raised; it was simply assumed that they were to blame.

Employers, on the other hand, saw the demand for punishment as a demand for revenge. Moreover, some of the relatives of the dead miners were union officials and employers believed that they were the victims of a union vendetta. For employers, Gretley was a terrible accident, for which no individual should be blamed.

The culpability puzzle

It was ultimately for the court to decide whether anyone was culpable. The mines inspectorate had provided Gretley mine with plans indicating where the old workings were. Unfortunately, the plans were wrong, out by 100 metres, and it was this error that ultimately led to disaster. The Gretley managers didn't check the accuracy of these plans. The court found that, given the well-known dangers of inrush from old workings, they should have. They hadn't done what reasonable mine managers would have done. They were therefore culpable and deserved to be fined heavily.

Many observers believed that the Department of Mineral Resources should have shared the blame, but the government's legal advice was that a prosecution against the department would fail. Hence, only the companies and their officers were brought to account.

The judgment of the court in the Gretley case raises various issues. With hindsight, it is easy to say that the managers should have checked the plans far more carefully. But is it fair to say that they should have had the foresight to do this, that they should have foreseen the possibility that the plans were in error, and that a reasonable manager would have foreseen this possibility? Just how culpable was their failure?

The issue is surprisingly complex and later chapters will examine in detail the reasoning that led to the court's conclusion. Moreover, not only did mine officials fail to check the accuracy of the plans, they failed to respond effectively when warning signs began to emerge in the days immediately prior to the accident. While the court did not blame them for this later failure, an earlier inquiry did. The findings of the earlier inquiry add an important dimension to the discussion of the culpability of the mine officials — a dimension that has been almost entirely overlooked in all of the commentary on the prosecution itself.

Outcome responsibility

There is a lot hanging on the answer to this question about culpability, quite apart from the fate of the individuals concerned. Suppose we conclude that the Gretley managers were not really to blame for what happened and therefore did not deserve to be convicted and fined. Does the law still have a role in holding individuals accountable for occupational disasters? Managers of large companies are frequently no more culpable when things go disastrously wrong than the managers at Gretley were. If the managers at Gretley did not deserve to be punished, then, probably, most managers in situations where industrial fatalities occur would not deserve to be punished. If so, the law (as currently formulated) can have no role in encouraging such managers to be more careful.

The problem is that the law (as currently formulated) must find fault before it can hold individuals accountable. I shall argue here that if we want to retain a secure role for the law in situations such as Gretley, we need to find a way to hold leaders accountable, even in the absence of fault. In making this argument, I draw on the English legal philosopher, Honoré, who shows that fault-based responsibility is only one type of responsibility. More fundamental, he says, is outcome responsibility, in which people are held responsible for outcomes, regardless of fault. To take a banal example: if I accidentally trip you, I owe you an apology and I am morally obliged to help you up, even if I was in no way at fault. The proposal, here, is that leaders be held responsible for what happens under their command, regardless of fault. It might not be appropriate to impose punitive sanctions in these circumstances, but courts could well impose other kinds of consequences that are designed to provide closure for relatives. At first sight, this proposal sounds far-fetched, but it turns out to have much in common with the principles of restorative justice, which aim to ease the pain of victims rather than impose additional pain on perpetrators. Indeed, the need for the legal system to adopt a restorative justice focus is one of the major conclusions arising from this case study.

The purposes of OHS law

Much of the public debate following the Gretley disaster centred on whether it was appropriate to hold mine officials accountable for the deaths of the four men. From a technical point of view, this involves a misconception of the purposes of the law under which they were prosecuted. Occupational health and safety law in Australia is not primarily aimed at holding people accountable after something has gone wrong. It is preventive, that is, it seeks to prevent things from going wrong in the first place. From this point of view, the aim of sentencing should not be to ensure that justice is done after the event, but to provide incentives to employers to take reasonable care or, to put it the other way round, to deter employer carelessness.

Employers argued that the Gretley defendants had taken reasonable care and that the real purpose of the law, to prevent harm, would not be served by convicting and punishing them. On this view, prosecuting the defendants after the event, in an attempt to ensure that justice was done, was really a misuse of the law.

Industrial manslaughter

Regardless of the legislative purpose, the public expectation is that the law will be used to bring employers to account when someone is killed. Not surprisingly, OHS legislation has not been particularly effective in doing this. This has led to calls for a new type of law, industrial manslaughter law, which envisages jailing individual employers in certain circumstances. So-called industrial manslaughter legislation has now been enacted in several Australian jurisdictions.

It is important to emphasise that such provisions have nothing in common with the provisions under which the Gretley managers were charged. They target individuals whose behaviour is grossly negligent or reckless — not words that could be used to describe the Gretley defendants. The sponsors of such legislation claim that it is aimed at the small minority of "rogue" employers — again, a word that is in no way applicable in the Gretley case. Senior executives are not at risk under industrial manslaughter legislation, unless of course they can fairly be described as rogues. The development of such legislation will be outlined in this book, both to demonstrate its limited applicability and to pinpoint the role that it can play in a broader framework of regulatory response to workplace fatalities.

Impact of the prosecution

The mining industry claimed that punishing people who were blameless, far from deterring carelessness, simply deterred good people from entering the industry. There were numerous public statements to this effect, some of them bordering on the hysterical. Very little evidence was adduced to support this claim. A book about Gretley could hardly avoid this issue, so I decided to carry out a small survey of mine managers to ascertain two things. The first was whether the prosecution had indeed had the effect of deterring people from taking positions of responsibility in the industry. The second purpose was to discover whether the prosecution had had any discernible effect in making managers more careful. The results of this survey are reported in the Appendix.

Mindful organisation and mindful leadership

Quite apart from the issues arising out of the prosecution, there is another reason that makes the Gretley case important: it highlights the issue of organisational mindfulness. As already alluded to, Gretley's risk management strategies failed in two ways. First, they failed at the outset to identify the risk that the plans might be in error. Second, in the days immediately preceding the disaster, there were warnings — admittedly ambiguous warnings — that something was amiss: increasing quantities of water were observed seeping from the mine face. Had these warnings been responded to effectively, the disaster would not have occurred.

The theory of high-reliability organisations suggests that these organisations are particularly alert to the significance of these kinds of warnings. They achieve safe and reliable operation in hazardous circumstances by mindfully attending to such warnings. One of the lessons of the Gretley disaster itself, as opposed to the prosecution, is the need for a style of operation that is more mindful of danger. If companies wish to move in the direction of high-reliability organisations, they must develop their systems for reporting and responding to the warning signs that invariably precede disaster. Interestingly, as we shall see, regulatory developments since Gretley, and more particularly since the Moura coal mine disaster in Queensland two years earlier, are prodding mining companies towards more mindful operation.

Mindful organisations require mindful leadership. This is the ultimate requirement for safe operation. This book therefore develops some ideas about just what mindful leadership means. Mindful leaders display "chronic unease" about the possibility that, unknown to them, something might be seriously awry in their organisation. Perhaps the most tangible way that this expresses itself is in their attitude to safety audits. I shall argue that mindful leaders do not look to audits for assurance that all is well; rather, they use audits to seek out the problems that they know are likely to be there. This is a crucial shift in viewpoint and, until leaders have made this shift, they cannot claim to be behaving mindfully.

The last chapter of this book draws together the lessons of the case study — for Gretley is rich in lessons. They cover both the role of the law in improving workplace safety and strategies for improving the mindfulness of organisations and their leaders. I argue, in particular, that outcome responsibility is precisely the kind of legal strategy most likely to encourage mindfulness.

Chapter 2

The demand for prosecution

The Gretley prosecution was unprecedented. When the authorities began contemplating a prosecution in 1998, there had never been a prosecution of a mining company under the NSW *Occupational Health and Safety Act* (the OHS Act) — despite the fact that prosecution had been possible since the Act's inception in 1983.[1] In contrast, prosecutions of companies in other industries were routine. Safety in the mining industry in NSW was overseen by a dedicated mines inspectorate and this inspectorate had a de facto non-prosecution policy. This long-standing policy of non-prosecution came to an end with Gretley, and two companies were charged and convicted under the OHS Act.[2]

The Gretley prosecution was also unprecedented in another way. A number of individual directors and managers in other industries had been prosecuted under the OHS Act for failing to ensure that their companies complied with the law. But these had mostly been very small companies, essentially one-person operations, where wrongdoing by this one person had led directly to the outcome of concern. There had been few, if any, cases in which managers or directors of larger companies had been held accountable for failing to ensure that their companies complied with the law (Foster, 2005). Gretley changed this as well. The prosecution targeted a number of people involved in management, among them, two mine managers — the current manager and a man who had managed Gretley mine two years earlier. The two managers, and a surveyor, were convicted and heavily fined. There were widespread fears beforehand that they might be sentenced to terms of imprisonment. By targeting managers in this way, the NSW Government was setting a precedent — not just for the mining industry, but for employers generally. The Gretley prosecution, then, was not just a case of catching up with practice in other industries; it represented a new trend of which other industries needed to take note. The case has therefore attracted interest well beyond the mining industry.

---

1   The mining industry was also governed by specific safety legislation — in the case of coal mining, the *Coal Mines Regulation Act 1982* (CMRA). Penalties under the CMRA were much lower than the penalties possible under the OHS Act. Prosecutions under the CMRA were virtually non-existent (see Staunton, J (1998), *Report of a formal investigation under section 98 of the Coal Mines Regulation Act, 1982*, p 687).

2   Newcastle Wallsend Coal Company Pty Ltd and Oakbridge Pty Ltd. The former was wholly owned by the latter. Once the issue of prosecution had been raised in the Gretley case, the authorities did begin to launch prosecutions in other cases and some of these resulted in convictions before the Gretley defendants were convicted.

The unprecedented nature of the Gretley prosecution provokes the questions: why did the authorities choose to act in this way, and why did they feel the need to abandon previous practice and hold companies and individuals accountable in the way that they did? This chapter offers an explanation.

In order to provide this explanation, the sequence of events that culminated in prosecution must be described in some detail. My purpose is to demonstrate the depth of feeling that the Gretley disaster engendered and the way in which that feeling was mobilised into what became a politically irresistible demand for prosecution. We shall see how pressure from the victims' families, the miners' union and, more broadly, the public combined to generate this outcome. In retrospect, it appears as an unstoppable process but, of course, to those involved, it was a matter of one step at a time, with no certainty at all about where it would all lead. Whatever else might be said about these events, they amounted to an extraordinary political phenomenon.

I shall also quote the words of those involved in the campaign at some length. I do so to provide a sense of what they were thinking and what their motives were. Later in this chapter, I shall seek to locate these motives within a more general framework of the purposes of punishment.

The context

Before embarking on this narrative, certain aspects of the context are worth outlining. As noted earlier, Gretley mine was on the outskirts of Newcastle, a city located at the mouth of the Hunter River. Coal mining has been the lifeblood of the Hunter Valley for more than a hundred years and numerous coal mining communities grew up in the area. The economy of the region is changing and the Hunter Valley is now synonymous with wine rather than coal in the minds of some people, but coal mining is still a major source of employment in the greater Newcastle region. Various Labor members of state parliament have connections with the mining communities in this area (or the mining communities in the Wollongong area) and are willing to represent the interests of coal miners in parliament.[3]

Because of Gretley's proximity to Newcastle, the Newcastle *Herald* printed regular stories as the Gretley events unfolded and wrote numerous editorials that were sympathetic to the relatives of those who had died. The newspaper both expressed and stimulated public concern about the tragedy and its aftermath.

---

3   Two such members spoke in the second reading of the Mines Legislation Amendment (Mines Safety) Bill, *Parliamentary Debates (Hansard)*, NSW Legislative Assembly, 21/10/98.

A further contextual aspect of the story concerns the role of the mining union, the Construction, Forestry, Mining and Energy Union (CFMEU). The union is influential with the Australian Labor Party (ALP) and the ALP was in government in the state of NSW throughout this period.[4] It was entirely sympathetic to union concerns.

Finally, one of the dead miners, Damon Murray, had important union connections. Both his father, Ian Murray, and his uncle, Peter Murray, were coal miners and prominent union officials. Ian, in particular, played a role in stimulating the government to action. Many observers have seen this family connection with the union leadership as a crucial part of the explanation of why government acted as it did. However, the matter is not quite so simple. The father of another of the dead miners was also frequently quoted in the media, urging action against the mining companies. He, too, was a coal miner but not a union official. In short, the families of the dead miners played an influential role, independently of direct connections to the union leadership.

Sequence of events

The accident at Gretley occurred in November 1996. In the preceding 20 months, there had been six deaths in the mining industry in NSW.[5] Moreover, shortly before the Gretley accident, a report was published on the Moura disaster in Queensland (that had occurred two years earlier) in which 11 men had died, and there was a need to consider how the findings of this report might be applied in NSW. Accordingly, just weeks before the Gretley disaster, the government had announced a review of mine safety in NSW — the Johnstone Review.[6]

The Gretley disaster gave added impetus to the Johnstone Review. But a specific inquiry into the causes of the accident was also necessary. The government had initially ordered a departmental investigation,[7] but the union believed that the Department of Mineral Resources was implicated and pushed for an independent judicial inquiry. The government accepted this advice and, two weeks after the accident, it announced that an independent inquiry would be set up. The head of this inquiry was later named as the former Chief Judge of the NSW District Court, James Staunton.[8] This was an important development. Inquiries often seek to draw lessons and not to apportion blame; certainly the Moura Inquiry two years earlier had consciously avoided allocating blame (Hopkins, 1999:4). An inquiry headed by a judge was more likely to allocate

---

4 I have had personal experience of this close connection. In 2000, I asked the union if it would organise a NSW launch of my book on the Moura mine disaster in Queensland. The union obliged. It organised for the book to be launched by the NSW Attorney-General in a functions room in the NSW Parliament.

5 The Minister for Resources, Mr Robert Martin, *Parliamentary Debates (Hansard)*, NSW Legislative Assembly, 20/11/96.

6 *Review of mine safety in NSW: report to the Minister for Mineral Resources and Fisheries*, ACiL, 14 March 1997 (Susan Johnston was the principal author of this report).

7 *The Herald*, 27/11/96. All references to *The Herald* are to the Newcastle *Herald*.

8 *The Herald*, 5/12/96.

blame and raise the possibility of prosecution, as indeed Judge Staunton did. The appointment of a judicial inquiry was thus an important first step along the path that ultimately led to prosecution.

The memorial service for the dead miners, and the events surrounding it, provide an insight into the political and emotional significance of the Gretley disaster and a demonstration of the influences that would eventually lead to the Gretley prosecution.[9] The service was held at the Sacred Heart Cathedral in Newcastle. The National President of the CFMEU delivered a eulogy in which he noted that more than 1,500 miners had lost their lives in the greater Newcastle area since 1847. "Once again, this industry has suddenly removed loved ones from our midst without even a handshake or a wave goodbye", he said. The state's Governor and the Premier also spoke to the congregation of more than 1,000. Significantly, in view of what was about to happen, the leader of the state opposition was also present. Outside the cathedral, the Premier said that the inquiry must satisfy the coal mining workforce. This was clearly a vital constituency.

By chance, the night before the memorial service, the NSW opposition had combined with independents in the Upper House of the NSW Parliament to cut workers compensation payments that would henceforth be paid to miners. Many who had attended the service were unaware of this development. When they discovered what had happened, they were infuriated, both by the decision itself but also by what was now seen as the hypocrisy of the Leader of the Opposition in attending the memorial service:

> "If he had any backbone, he would have made his stance on miners' compensation payments public before inviting himself to the service", Mr Ian Murray said. "His timing was atrocious, if anything he is going to lose because of it."[10]

The Leader of the Opposition was clearly caught out. He declined to respond to Ian Murray's statements, saying that he had "only the deepest, most profound sympathy for those families". Miners called an immediate and indefinite strike. The Leader of the Opposition was forced into what one newspaper report called an "embarrassing backdown" and the independents announced that they would reverse their vote. The disaster was already having political consequences.

The Staunton Inquiry reported in June 1998.[11] It was very critical of both the company that operated the Gretley mine and its management, and it recommended that the matter be referred to the Crown Solicitor to determine

---

9   *The Sydney Morning Herald*, 29/11/96, 30/11/96; *The Daily Telegraph*, 28/11/96, 30/11/96.

10   *The Daily Telegraph*, 30/11/96.

11   Staunton, J (1998), *Report of a formal investigation under section 98 of the Coal Mines Regulation Act, 1982.*

whether there were grounds for prosecution under the NSW OHS Act. The government announced that it would implement all of the recommendations of the inquiry[12] — and it could hardly have exempted the recommendation for prosecution from this blanket commitment.

The Staunton Inquiry was also highly critical of the mines inspectorate's policy of non-prosecution to that time. It noted that, since April 1990, there had been 33 deaths in the coal mining industry alone in NSW. It examined documents in relation to these fatalities and concluded that "almost without exception, the documents revealed significant breaches of the Act, and Regulations or an obvious failure to take reasonable care for the safety of workmen".[13] And yet, it noted, in not one of these cases had a mining company been prosecuted. Indeed, "in most instances there was either no consideration, or insufficient consideration, given to the prosecution of those responsible".

The Chief Inspector of Mines gave evidence that his preferred response to fatalities was not to prosecute but to carry out a "system safety accident investigation", with a view to making recommendations to avoid a recurrence. "My assessment of the situation", he said, "was that we would gain more from proceeding the way we did than launching an OHS action".[14]

The inquiry rejected this view and concluded: ". . . there is an urgent need for change . . . There should, in appropriate cases, be prosecutions. It is important that the industry knows when it makes decisions that it is accountable under law for those decisions."[15]

It should be mentioned that this general policy of non-prosecution was a feature of regulatory regimes for the mining sector in other states as well. Most conspicuously, there was no prosecution following the Moura tragedy in Queensland two years earlier, even though the Moura Inquiry had expressed the view that the events leading to that disaster involved "management neglect and non-decision which must never be repeated in the coal mining industry" (Hopkins, 1999:4).

The Staunton Inquiry's stringent criticism of the policy of non-prosecution in NSW gave added impetus to its recommendation that consideration be given to prosecution in the Gretley case.

The publication of the Staunton Inquiry in June 1998 also stimulated public demands for prosecution. An editorial in the Newcastle *Herald* was highly critical of the policy of non-prosecution revealed in the inquiry. It ventured the view that "if a prosecutions policy had been in place before the Gretley accident, the errors which led to the accident may not have occurred".[16] The union joined in and announced that, if necessary, it would launch a prosecution itself. This was no

---

12   *Parliamentary Debates (Hansard)*, NSW Legislative Assembly, 21/10/98.

13   Staunton, op cit, pp 687, 696, 697.

14   Ibid, pp 697, 698.

15   Ibid, p 707.

16   *The Herald*, 11/7/98.

idle threat. Section 48(1) of the NSW OHS Act allowed unions to initiate prosecutions, although the provision has only rarely been used.[17] The National Secretary said that the findings of the Staunton Inquiry "vindicated the stand taken by the families of the deceased and the union". He went on:

> ". . . we are sick and tired of the carnage in the mining industry and we believe it is time that management and companies were called to account for the consequences of their negligence. Our union holds the names of 2,280 coal mine workers killed in NSW mines and, in all the time since the first fatality, not one person or company has ever been prosecuted for any of these deaths in the history of NSW."[18]

Union representatives also noted that Judge Staunton's recommendation referred only to the possibility of prosecuting companies. They were adamant that government should go beyond this in two respects. It should consider prosecuting individuals in management who may have been at fault, and it should consider prosecuting the department which had supplied the incorrect plans to the mine. A spokesman for the families reiterated the need to prosecute individuals. "People need to be held accountable for their actions", he said.[19]

Six months after the release of the Staunton Inquiry, the Crown Solicitor had still not made a decision on what, if any, criminal charges would be laid. The union was scathing. "The union finds it incomprehensible that the Crown Solicitor has taken this long to determine whether or not there should be prosecution in what seems to us a fairly open and shut case", a spokesman said. "The Crown Solicitor has got to get off his arse, it's as simple as that." The delay was "unacceptable" and "a disgrace". He went on, "we would hope the government recognises that these families [of the dead miners] have been through more than enough already without this delay with a prosecution". He threatened again that if the government did not prosecute, the union would. A spokesman for the Minister for Mineral Resources and for the Attorney-General responded sympathetically. He said that they appreciated the concern of the families and understood their frustration. He explained that the Crown needed to be certain of its case before proceeding.[20]

In the latter part of 1999, the families announced that they would sue the companies and the Department of Mineral Resources for nervous shock. They were particularly incensed by the department's stated belief that it was immune from prosecution because it did not have a legal duty of care towards the dead miners.[21]

---

17   *The Herald*, 8/7/98, 5/8/98.

18   CFMEU press release, 7/7/98.

19   *The Herald*, 9/7/98.

20   *The Herald*, 7/1/99; AAP, 7/1/99.

21   *The Herald*, 19/9/99; *The Sydney Morning Herald*, 18/9/99.

Towards the end of the year, just prior to the third anniversary of the disaster, the Attorney-General announced that charges would be laid. "The government is determined to see that justice is done", he said. But he did not specify precisely who would be targeted.[22] That was revealed several months later in April 2000, when two companies and eight individuals, including two mine managers, were charged. The Department of Mineral Resources, which had provided the erroneous maps, was not charged because, the government said, its legal advice was that such a charge would probably fail. The union was furious, again threatening that it might seek to prosecute the department. According to a spokesman, "the sad thing about this is that politics has taken precedence over the memories of the four men who lost their lives".[23]

At this point, the state opposition sensed an opportunity. It accused the government of a cover-up. It cited another legal opinion that a prosecution against the department might succeed and it demanded that the government table its own legal opinion. A spokesman explained that the opposition "wanted due process for both the families and the miners". The government responded to this pressure by forwarding the evidence to the Director of Public Prosecutions (DPP) for a third opinion. The opposition spokesman described the government's backdown as a victory for due process and a comfort to the relatives of the four Gretley victims. "I phoned the father of one of those young miners and it was a very special moment", he said. "I am so pleased it has been able to bring some solace to them." However, the opposition's victory was hollow. Two months later, the DPP confirmed that a prosecution against the department would fail.[24]

Then began an extensive period of legal manoeuvring, of which the union and families were highly critical. It seems likely that this legal manoeuvring included an attempt by the parent company, Xstrata, to plea bargain, that is, to plead guilty in return for certain concessions. The company was very keen to protect its individual employees from liability, with the Xstrata CEO saying at one stage: "I believe there's an absolute obligation on the company to ensure that these people are given the very best opportunity to avoid being convicted of a criminal charge associated with the incident."[25] It is therefore probable that the company offered to plead guilty if the charges against the individuals were dropped. This was a crucial issue. Most OHS charges are settled by a plea of guilty, which saves everyone the time and expense of going to court. But had the charges against the individuals been dropped, the political fallout would have been enormous. The prosecution could not afford to accept such a bargain — and it probably could not have accepted any other bargain in the politically charged atmosphere that

---

22   *The Herald*, 12/11/99, 13/11/99.

23   *The Herald*, 18/4/2000; AAP, 17/4/2000.

24   AAP, 5/5/2000, 2/6/2000; *The Herald*, 3/6/2000, 28/7/2000.

25   *The Herald*, 10/8/04.

prevailed. It therefore rejected the offer.[26] The decision by the prosecution not to accept a plea of guilty must be seen as a crucial step along the path leading to the unprecedented convictions that followed.

This period of legal manoeuvring culminated, 18 months after the charges were laid, in formal not guilty pleas. A spokesman for the families said that he was "over the moon" about this outcome. It meant, he said, that the matter would go to trial and every piece of evidence would be aired. Guilty pleas would have seen the matter "squared away with the minimum of fuss", but the families would have been denied the answers that they so desperately craved, he said.[27]

The legal process dragged on into 2003, with public demands from time to time that it be accelerated. An editorial in the Newcastle *Herald* expressed this sentiment. It was titled, "The Agony of Gretley". It said, among other things: "the continuing [legal] delays ... are deplorable. The grieving families of the four men who died have been denied closure"; the trial judge "should halt the stalling" by the defence; and "enough is enough".[28] The Attorney-General went on record again, saying that "the government is determined to see that justice is done".[29]

Then, in November 2003, the defence dropped a bombshell. It challenged the legality of the whole prosecution by pointing out that the wrong minister had signed the documents authorising the prosecution. The union responded that, if the case was dismissed on this basis, it would shut down the nation's coal industry. A spokesman for the families said that the legal challenge "continued a shameful campaign by the mining company to avoid facing up to its responsibilities". The government quickly announced that it would legislate to allow the prosecution to be retrospectively authorised, thereby averting the threatened strike.[30] Clearly, the momentum was now enormous.

Finally, in August 2004, came the verdicts: two corporate entities and three individuals convicted. An editorial in the Newcastle *Herald* noted that the convictions "hopefully mean that the families of the four miners killed ... will

---

26 Whether or not an attempt to plea bargain was made during this period, the record shows that such an offer was made and rejected after the trial commenced ([2005] NSWIRComm 31, para 368, 369). See also comments by Marks J in *Newcastle Wallsend Coal Company Pty Ltd & Ors v Inspector McMartin* [2006] NSWIRComm 339 (5 December 2006), para 744-746.

27 *The Herald*, 15/11/02, 7/12/02.

28 *The Herald*, 15/7/03.

29 *The Herald*, 15/7/03, p 1.

30 *The Herald*, 21/11/03, 22/11/03, 25/11/03, 26/11/03; Australian Broadcasting Corporation (ABC), 21/11/03, 29/11/03.

soon be able to have less troubled sleep". It declared that "[the] findings are a historic first for this state's coal industry. It has long been a bone of contention for trade unions that mining deaths have been viewed as accidents, with no company or management employee held to be responsible". Any appeal by the defendants against the decision "would be disappointing", it said.[31] With sentencing yet to come, union representatives said that "those responsible deserve a hefty fine"; indeed, "they should face the maximum penalties".[32]

The union was not, however, satisfied with the existing state of the law. It urged the government to introduce tougher manslaughter legislation so that mine managers and companies could be more effectively prosecuted in the future.[33]

In September 2004, following the deaths of three more miners in separate incidents in the preceding 12 months, the government established another high-level review into mine safety — the Wran Review.[34] This was yet another indication of the political sensitivity of the issue of mine safety in NSW.

In January 2005, Xstrata dropped another bombshell — as far as the unions and families were concerned. It announced that it would appeal the convictions. A Sydney newspaper, *The Daily Telegraph*, noted that the decision would re-open painful wounds. It quoted a family spokesman as saying, ". . . to have the matter raised again after it was all said and done is just heartbreaking".[35]

In March 2005, the Gretley defendants were sentenced, with fines totalling about $1.5 million. This was a record fine under the OHS Act in NSW. But the knowledge that the defendants were appealing the convictions undermined the outcome for the union and the families. According to one family member, the fines meant little because the company had not accepted responsibility for the tragedy. Unless the company accepted responsibility, the four men would have "died in vain", he said.[36] A few days later, the union held a rally in Sydney to protest against the appeal. The rally was about "respect for the dead". As the president of the CFMEU put it, "You know who doesn't respect the dead, those who want to wriggle out of prosecutions". Mine workers resolved to shut down the industry if the appeal succeeded. A family spokesman described Xstrata as

---

31   *The Herald*, 10/8/04.

32   ABC, 15/11/04.

33   ABC, 5/11/04.

34   Wran, N and McClelland, J, *NSW mine safety review*, Report to The Hon Kerry Hickey MP, Minister for Mineral Resources, February 2005.

35   ABC, 29/1/05, 31/1/05.

36   ABC, 11/3/05; *The Herald*, 12/3/05.

"morally bankrupt" and a "corporate thug" because of its refusal to provide peace and closure to the families of the dead miners.[37] For the record, it should be noted that the appeal, when it was finally determined, was unsuccessful in nearly all respects.[38]

Factors leading to prosecution

It can be seen, then, that a number of circumstances converged to produce the Gretley prosecution. Unlike the Moura disaster which occurred in outback Queensland, far from any centres of population, Gretley was on the outskirts of a major city (Newcastle) and public sympathy was therefore readily mobilised. Furthermore, Gretley was located in the heart of a large and populous coal mining region that routinely elected to the state parliament people with close connections to the mining workforce. These parliamentary representatives took a particular interest in events and reinforced the government's commitment to take action.

The CFMEU exerted enormous influence over the government of the day (the ALP). This was not just influence stemming from a convergence of political views. It stemmed in part from the union's capacity to close down the mining industry. A strike on an issue as emotional as justice for the miners who had died would have been very difficult for any government to manage.

The public statements from union officials and families of the dead miners were a crucial element in the mix. The anguish of the families and the emotional, yet articulate, way in which they expressed it appeared to carry all before it. Xstrata had no answer for their stream of commentary on events.

The decision to set up a judicial inquiry was also pivotal. Not only did the inquiry recommend prosecution, but it was also scathing about the existing policy of non-prosecution. The government could hardly fail to respond.

---

37   ABC, 14/3/05; *The Herald*, 15/3/05.

38   *Newcastle Wallsend Coal Company Pty Ltd & Ors v Inspector McMartin* [2006] NSWIRComm 339 (5 December 2006). The convictions and sentences of the two companies concerned were upheld. The conviction and sentence of the manager at the time of the accident were upheld. The conviction of the surveyor was quashed on the grounds that he was not a person concerned in the management of the corporation. In relation to the former manager, the appeal court upheld convictions on certain counts and quashed the convictions on others, on what can reasonably be described as technicalities. It was therefore necessary to re-sentence this defendant. In so doing, the appeal court chose to apply section 10 of the *Crimes (Sentencing Procedure) Act 1999*, which provides that, when a "court finds a person guilty of an offence", it may dismiss the charges without conviction. It applied this provision "not without some real misgivings" (para 613). Thus, although the appeal court effectively reversed the outcome for this individual, it found that he was guilty on at least some of the counts on which he was originally convicted. In short, the appeal court agreed with the lower court that both managers were guilty. This is not an outcome that company managers in general can take comfort from. These were majority findings. The dissenting judge agreed with the majority that the companies were guilty on some counts, and that both managers were guilty on some counts. Again, there is not much comfort here for managers in general.

Moreover, although according to custom, the NSW OHS Act had not been applied in the mining industry, there was no legal impediment to doing so. All that was required was the political will — which was there in abundance. In contrast, BHP, the operator of the Moura mine in Queensland, was effectively immune from prosecution under the legislation that existed at the time.

Finally, fatalities continued to occur throughout this period, reminding everyone of just what was at stake. Two high-level safety inquiries were set up independently of the unfolding Gretley events, but they ran concurrently with them and contributed to the pressure.

Gretley was certainly a landmark prosecution. It occurred because a particular set of circumstances combined to produce the necessary political impetus. Such a combination of circumstances is unusual, but not unique. A similar set of circumstances in Victoria produced the unprecedented prosecution of Esso Australia in 1999, following the gas plant accident at Longford (Hopkins, 2000, ch 1). Accidents involving multiple fatalities do not invariably generate this kind of response. It would be interesting to explore in more detail why some accidents do and some don't, but that is beyond the scope of this book.

The purposes of punishment

The previous discussion provides an account of the political circumstances that led to prosecution.. The aim was to explain why such an unprecedented prosecution occurred. In the remainder of this chapter, the focus shifts somewhat. I want to explore the reasons given by the union, the families and others for demanding prosecution. How, in their minds, was it justified? This question can be most effectively answered by beginning with a general account of the purposes of punishment.

We can distinguish two broad categories of justifications for punishment or, more generally, the criminal justice system: consequentialist and desert-based justifications. Let us consider, first, consequentialist justifications, that is, justifications in terms of the *consequences* that punishment, or the criminal justice system, seeks to achieve. Deterrence and rehabilitation are well-known consequentialist justifications.[39] Of course, as soon as such consequences are advanced as justifications, the question arises as to whether the criminal justice system in fact achieves these outcomes. This is an empirical question which can only be answered by painstaking research. Consequentialist justifications depend on the outcome of this research; they only make sense if research demonstrates that such effects indeed occur.

---

39 For a fuller discussion, see Bagaric (2001). Chapter 6 in Gunningham and Johnstone (1999) provides a useful account of the purposes of punishment in the context of OHS offences. Honderich (1971) is a classic and useful discussion of the topic.

It is important to distinguish between two kinds of deterrent effects of punishment: the effect on the individual who is punished, and the effect on others who become aware of the punishment. The first is called specific deterrence and the second, general deterrence. It is possible for a punishment to have one but not the other effect. For instance, it is a plausible hypothesis that the threat of being fined for speeding has little effect on us until we are caught and fined, after which we drive in a more law-abiding way. If research demonstrated this to be the case, we would conclude that traffic fines have a specific deterrent effect but little or no general deterrent effect. This distinction between specific and general deterrence will be of considerable significance in later chapters. The sentencing judge made reference to these two different types of deterrence, justifying the sentence, in part, in terms of general deterrence. I shall also report on a small study that was carried out as part of this project to determine whether the Gretley prosecution in fact had any deterrent effects, either specific or general (see the Appendix).

Another consequentialist justification (mentioned here because I shall touch on it again below) is incapacitation. If the offender is locked away in prison, he or she is incapable of further crime, at least against non-incarcerated citizens, for the duration of the sentence. A relatively large amount of crime is committed by a relatively small number of people, and there is some evidence that, if these people can be apprehended and incarcerated for significant periods of time, there is an overall reduction in the crime rate (Sheley, 2000:609-611). In short, crime rates can be reduced by incapacitating recidivist offenders. The ultimate incapacitation strategy is capital punishment.

The second type of justification — desert — is absolute in nature: offenders must be punished because they *deserve* punishment. More generally, the purpose of the criminal justice system is to ensure that offenders receive their just deserts. This is often referred to as retribution.[40] The question of whether punishment has any beneficial consequences, such as deterrence or rehabilitation, is irrelevant to the strict retributivist. Historically, retribution has been the most fundamental of the justifications for punishment. A familiar expression of this idea is the old adage: an eye for an eye and a tooth for a tooth.

There are obviously close similarities between retribution and revenge. But advocates of retribution deny that they are the same: retribution, they say, is a matter of principle, while revenge is simply a human motive, and a rather base one at that (Honderich, 1971:15, 42). At first sight, this looks like splitting hairs, but it would be a mistake to dismiss retribution as merely revenge in disguise. Importantly, the concept of desert not only justifies punishing the guilty — it prohibits punishing the innocent. Moreover, it limits punishment of the guilty to that which they deserve.[41] Let us dwell on this point for a moment. It may be that

---

40   Such a justification is also called deontological or axiomatic, that is, underived from any prior consideration. See Braithwaite and Pettit (1992:33).

41   Braithwaite and Pettit (1992:35) describe this as negative retributivism.

the best way to protect the public from persistent petty thieves is to lock them up for life, but most of us would want to argue that persistent petty thieves do not *deserve* such a fate. Similarly, even if it could be shown that the death penalty effectively deterred pickpocketing, we would surely object that pickpockets do not *deserve* the death penalty. Finally, it may be that the most effective way to deter some offenders is to inflict harm on their spouses or children. The principle of desert rules this out in a liberal society; family members do not *deserve* to be punished for a crime that they did not commit.[42] In short, the principle of desert serves to limit the excesses to which consequentialist justifications might lead. If we use the principle of desert to limit the extent of punishment in this way, it becomes more difficult to dismiss a desert-based argument in favour of punishment as mere revenge.

Those who argue for retribution sometimes revert to consequentialist arguments without fully recognising it. For instance, Lord Denning has said that "the ultimate justification of any punishment is not that it is a deterrent but that it is the emphatic denunciation by the community of a crime" (Braithwaite and Pettit, 1992:160). We are entitled to ask of Lord Denning, why is this denunciation important? Two kinds of consequentialist answers might be given. The first is that denouncing the crime or the criminal makes it less likely that the offender will reoffend or that others will offend in the same way. It does so by reinforcing the moral significance of the law which has been broken. This is a moral-educative effect and, in so far as punishment has this effect, it contributes to crime prevention. The second consequence has nothing to do with crime prevention. The community demands that a strong statement of disapproval be made and punishment satisfies that demand. There is an important implication here. If we take account of the community need for offenders to get their just deserts, even the retributivist justification becomes consequentialist. In other words, giving offenders their just deserts can be justified in terms of the consequences that it has for the community, independently of any presumed consequences for crime prevention.

We can develop this consequentialist line of argument by asking: why is it desirable that community demands for retribution be satisfied? Various answers are possible. For instance, if the criminal justice system did not respond in some degree to community demands for retribution, the citizenry might take the law into their own hands, with lynch law the result. The justice system must therefore respond in some way to community demands for retribution in order to maintain its own legitimacy and to avoid such anarchic outcomes.

Retribution has another important consequence for the community. Where people have died as the result of an offence, relatives often feel that they cannot move on with their lives until some form of retribution has been exacted from offenders. The interests of relatives in these circumstances are a legitimate

---

42   In some traditional societies, guilt may extend to the whole family or even the whole tribe.

concern of the criminal justice system, and retribution is sometimes justified by the need to provide relatives with "closure". Here, again, the retributivist justification becomes consequentialist in nature.

A particularly eloquent statement of this need for closure was made by the wife of one of the miners killed in the Moura disaster, "You pick up the pieces of your life and you stuff 'em back into a container of normality and you try to stuff 'em back into a cocoon of mundaneness and hope that one day it will all be together again" (Hopkins, 1999:5).

Not all retributivist arguments are consequentialist. Some retributivists insist that punishment is an absolute good, independently of any consequences that it may have for relevant audiences (Braithwaite and Pettit, 1992:160). But the point I want to stress here is that retribution, that is, giving people their just deserts, *can* be justified in terms of the needs of victims, their families and the wider community, in ways that have nothing to do with crime prevention. The interests which relatives may have in retribution are not often considered by those who theorise about the justifications of punishment but, in practice, the criminal justice system takes these interests explicitly into account — as we shall see below.

The purposes of punishment: a legislative statement

It is interesting to juxtapose the preceding, rather abstract, account of the purposes of punishment with the statement of the purposes of punishment contained in the NSW *Crimes (Sentencing Procedure) Act 1999*. I do so for two reasons. First, this legislative statement of purpose guided the judge in the Gretley case (as I shall show in a later chapter). Second, and more importantly in the present context, I want to show how one legislature, in this case the NSW legislature, invokes the various justifications discussed above and, in particular, invokes retribution — *both* for absolute reasons *and* in order to achieve various consequences. Section 3A of the Act states that:

Crimes (Sentencing Procedure) Act 1999, section 3A

The purposes for which a court may impose a sentence on an offender are as follows:

(a) to ensure that the offender is adequately punished for the offence,

(b) to prevent crime by deterring the offender and other persons from committing similar offences,

(c) to protect the community from the offender,

(d) to promote the rehabilitation of the offender,

(e) to make the offender accountable for his or her actions,

(f) to denounce the conduct of the offender,

(g) to recognise the harm done to the victim of the crime and the community.

Let us consider these purposes in turn. Purpose (a), ensuring adequate punishment, is a straight-out retributivist justification. The aim is to ensure that the punishment fits the crime, precisely in the way that the absolute retributivist would want. There is no crime prevention element here, nor is there any suggestion that adequate punishment is necessary to satisfy a community need.

Purposes (b), (c) and (d) are all about crime prevention. Purpose (b) aims to prevent crime by deterrence (both specific deterrence of the offender and general deterrence of others). Purpose (c), protecting the community from the offender, is most obviously achieved by locking the offender away. This is the incapacitative strategy mentioned above. Purpose (d), rehabilitating the offender, is obviously preventive.

Purpose (e), making the offender accountable for his or her actions, is rather more ambiguous from the present point of view. Why should people be held accountable? Perhaps the intention is that that they and others will be encouraged to take their legal obligations more seriously in the future. On this reading, holding people accountable is desirable for the purposes of crime prevention. Alternatively, purpose (e) can be read as another statement of the principle of retribution: those who break the law *deserve* to be held to account, regardless of whether this serves any crime prevention purpose. Perhaps the legislature had both these purposes in mind.

Purpose (f) is to denounce the conduct of the offender. As discussed above, the purpose of denunciation may be, in part, crime prevention but it also contains a retributivist element that is aimed at satisfying a community need.

Finally, purpose (g) is to recognise the harm done to the victim and the community. Clearly, what is envisaged here is retribution — imposing just deserts on the offender — in order to satisfy a community need.

This analysis reveals that, in addition to prevention, a major goal of sentencing is indeed retribution, that is, ensuring offenders get their just deserts. The legislature has instructed the courts that retribution is not only inherently desirable but it is also necessary in order to satisfy various community needs. It is justified both on grounds of absolute principle and in terms of the consequences that it is intended to achieve.

The purposes of the Gretley prosecution

The sections above have laid out a framework for thinking about the purposes of punishment. We can now return to the Gretley case and consider in more detail the comments made by the various parties about the need for prosecution. Of course, very few of those who made statements about the desirability or otherwise of prosecution had a clear idea of the range of justifications sketched above, and many of their comments will need to be interpreted in order to draw out their significance. But it is important to engage in this interpretative exercise in order to make sense of what was said.

Some of the positions advanced were unambiguously consequentialist and focused particularly on the preventive effects of prosecution. Recall the comment from the Newcastle *Herald* that "if a prosecutions policy had been in place before the Gretley accident, the errors which led to the accident may not have occurred". This was an explicitly consequentialist argument — prosecution was desirable for crime prevention purposes. The position of the Chief Inspector of Mines was similarly consequentialist, although diametrically opposed to that of the newspaper. His view was that more was to be gained from a "system safety accident investigation" than from a prosecution. The question in his mind was whether prosecution was the most effective way to encourage companies to take greater care, and his answer, generally, was no. Educational and informational strategies would, he believed, be more effective. The fact that interested parties can argue both for and against prosecution in terms of crime prevention highlights the need for careful empirical research to determine what the effects of prosecution really are.[43]

One of the most frequently expressed arguments for prosecution was that the companies and individuals needed to be held responsible or accountable. This was Judge Staunton's view: "[Industry must know] when it makes decisions . . . it is accountable under law for those decisions." As noted above, accountability can be argued for on crime prevention grounds, specifically the need for deterrence, and also as a matter of desert. Probably the judge had both of these purposes in mind.[44]

This ambiguity runs through most of the comments about accountability. Recall the union comment, "we are sick and tired of the carnage in the mining industry and we believe it is time that management and companies were called to account for the consequences of their negligence". This comment embodies a consequentialist belief that holding people to account may bring an end to fatalities, but there is also a strong suggestion that managers and their companies *deserve* to be prosecuted for their negligence.

Perhaps the clearest indication of the retributivist element in the demand that people be held to account came from one of the family representatives, "those responsible deserve a hefty fine". There is no suggestion here of any benefits that might stem from identifying and punishing those who are responsible, it is simply that they *deserve* to be punished.

---

43 I shall not review here the empirical literature on the deterrent effects of punishment. The Appendix to this book reports on an empirical study of the deterrent effects of the Gretley prosecution.

44 Judges who speak of the need for deterrence when they pass sentence seldom raise the empirical question of whether the punishments that they impose in fact have any deterrent effect, and it is usually clear from reading their comments that retribution is uppermost in their minds. See Hopkins (2002:34, fn 5).

A third prominent theme was the need for justice. "The government is determined to see that justice is done", a spokesman said at one point.[45] This comment seems remote from any consequentialist considerations. It is a pure retributivist statement: if justice requires that offenders be punished, they will be.

The strongest theme running through public commentary was the expectation that punishment would, in some way, heal the wounds suffered by the families of the victims and provide them with some sense of closure. So, when the delays in launching a prosecution seemed interminable, a family member commented, "we would hope the government recognises that these families [of the dead miners] have been through more than enough already without this delay with a prosecution". When the convictions were finally announced, the Newcastle *Herald* declared that the families would "soon be able to have less troubled sleep". Xstrata's announcement that it would appeal the verdict provoked numerous comments (in the press) from family members, such as:

> "After more than eight years of hell, today should have been the end of the matter but [the] appeal means that this has a long way to go."[46]

> "[This will] cause more distress for the victim's families . . . to have the matter raised again after it was all said and done is just heartbreaking . . . they should cop the punishment and let us get on with life."[47]

> The appeal involved a "refusal to provide peace and closure to the families of the dead miners".[48]

These comments all reflect the benefits of retribution for the bereaved.

Some comments appeared almost to suggest that prosecution was in the interest of those who had died. The failure to prosecute the Department of Mineral Resources meant "that politics has taken precedence over the memories of the four men who lost their lives", and the appeal against the convictions showed a lack of "respect for the dead". Of course, the dead themselves can have no interest in prosecution; it is those who remain behind who have the memories and who want respect for their dead. These comments are variations on the theme that retribution provides some form of compensation to those who remain behind.

It is clear, then, that the demands for prosecution were based on a wide array of justifications, both absolute and consequentialist. On the whole, the parties did not make their arguments explicitly in these terms, but their comments can easily be interpreted in this way. One of the advantages of this analysis is that it makes the demand for prosecution rather more intelligible. Coal mine companies expressed the view that they had been the victims of a "vendetta" pursued by

---

45   *The Herald,* 12/11/99.

46   *The Herald,* 12/3/05.

47   *The Daily Telegraph,* 31/1/05.

48   ABC, 14/3/05; *The Herald,* 15/3/05.

the families and the union, but the demand for retribution cannot be dismissed in this way. Justice, accountability and closure for relatives are all ideas that contain elements of retribution, and they command widespread support. Understanding the philosophical basis of the campaign by the families and the union contributes to our understanding of why it succeeded.

The idea of retribution normally pre-supposes fault, culpability and blameworthiness. The principle of imposing suffering by way of payback has a certain plausibility if the initial action was in some way blameworthy. Otherwise it is merely scapegoating. The Staunton Inquiry implicitly recommended that a company should be prosecuted, but there was no such recommendation, even implicitly, in relation to individuals. Those who demanded that individuals be prosecuted simply assumed that they were sufficiently at fault to warrant being made to suffer. In the end, the court agreed that they were, and it imposed a second round of suffering in response to the first. Subsequent chapters will explore the court's conclusion that these individuals were sufficiently culpable to warrant being made to suffer in this way.

Restorative justice: a first mention

There is an important implication of the earlier analysis with which I shall finish this chapter. As we have seen, one of main arguments for prosecution was that retribution would provide some degree of closure for relatives. As soon as retribution is justified in consequentialist terms such as these, the way is open to consider whether the desired consequences might be achieved by other means that do not involve the infliction of pain.

There are indeed other means. Work in the United States has shown that, when corporate scandals give rise to public outrage and consequent demands for retribution, the outrage can be significantly diminished — and with it the demand for retribution — when corporate leaders apologise extensively and sincerely and do their utmost to rectify the harm that has been done.[49] Moreover, the restorative justice movement (about which I shall say more in a later chapter) seeks repair and closure for victims, rather than pain for offenders. The outcome may involve reparation by the offender and an apology. I am not suggesting here that a mere apology would have satisfied the demand for retribution in the Gretley case; but a fulsome apology, accompanied by tangible evidence of its sincerity, might have assuaged that demand somewhat. The point is that understanding the purposes of punishment can suggest more humane and possibly more effective ways in which those purposes can be met.

---

49   See the work of Peter Sandman at www.psandman.com.

Chapter 3

The first failure: reliance on faulty plans

Inrush of water from old workings is a well-known mining hazard. There were at least 10 inrush events in United Kingdom coal mines in the 19th and 20th centuries, resulting in more than 300 deaths.[1] The need for precautions is therefore well understood. How could these precautions have failed at Gretley when old workings were known to be in the vicinity? This chapter examines management's failure to verify the accuracy of the plans being used. It explores in detail the court's findings about culpability, in particular, the culpability of the individuals who were convicted. The chapter that follows will deal with management's failure to respond effectively to the warning signs once they began to appear.

Several years before the Gretley inrush, management had obtained from the Department of Mineral Resources some old plans that purported to indicate the location of the earlier workings.[2] Unfortunately, the plans were wrong because they indicated that the old workings were 100 metres further away than they actually were. Initially, mining at Gretley was quite remote from the old workings. However, in 1994, Gretley management began planning a new section of the mine that would take it close to the old workings.[3] Using the department's plans, Gretley management drew up its own plan showing where it believed the old workings to be and where it proposed to mine. The mine surveyor at the time assumed that the plans provided by the department were accurate and the manager, relying on his surveyor, made the same assumption. At the time of the tragedy (two years later), the mine had a new surveyor and a new manager, and both men assumed that the plans were accurate, relying on the judgments of their predecessors. Indeed, the new surveyor certified the accuracy of the plans in writing, based simply on the fact that they had been so certified by the previous surveyor.[4] Both managers had also implicitly certified the accuracy of the plans in writing.[5]

---

1 I p 90 ("I" refers to the inquiry by Judge James Staunton (1998), *Report of a formal investigation under section 98 of the Coal Mines Regulation Act, 1982*).

2 I p 271.

3 The application to mine was dated September 1994. Plans of the old workings included in the application were dated 31 July 1994 and were signed by the mine manager and the mine surveyor. J 395, 396 ("J" refers to the judgment handed down in *McMartin v Newcastle Wallsend Coal Company Pty Ltd & Ors* [2004] NSWIRComm 202 (9 August 2004). Numbers following a "J" refer to paragraphs).

4 J 968.

5 J 963.

The faulty plans were produced by a departmental draftsman for another purpose in 1980.[6] So it was that a drafting error, made long before the Gretley mine began operations, culminated in disaster 16 years later. This is a classic example of what Reason (1997) has called a latent error — an error that lies dormant for many years before, in conjunction with new circumstances, its catastrophic potential is realised. As one observer noted, the plans "sat like a loaded gun in the archives",[7] waiting to be fired.

The operating company and its owning company[8] were charged and found guilty under the NSW OHS Act for failing to ensure the safety of employees, so far as reasonably practicable.[9] The companies were aghast that they should be blamed in this way. They argued vigorously that they had been entitled to rely on the accuracy of the plans and that they were therefore not responsible for the tragedy. Judge Patricia Staunton found, however, that management should not have relied on the plans provided and should have sought to verify their accuracy independently. Her reasoning on this point is worth quoting:

> "The defendants submit that the cause of the inrush was [the department's drafting error] . . . I fundamentally disagree. The cause of the inrush, in my view, was the failure by the defendants to properly research the location and the extent of . . . the old workings."[10]

This is a puzzling statement. Had the department not made the error it did, the accident would not have happened. Had the defendants researched the location of the old workings more carefully, the accident would not have happened. From this point of view, if one is a cause, so is the other. Clearly, therefore, the judge's statement is not to be understood in this way. Instead, by placing the causal emphasis where she did, she was making a statement about the relative culpability of the parties.[11] Her thinking was as follows:

> "There is no doctrine of implied infallibility to be applied to the information, documentary or otherwise, given out by any government department. While it is reasonable to presume that such information would generally be correct, that in no way removes the defendant's independent obligation to ensure the accuracy of the information."[12]

---

6   J 389.

7   I p 711.

8   J 280.

9   J 824.

10   J 805.

11   The law often restricts the idea of cause in this way, eg Mason J in *March v Stramare Pty Ltd* (1991) 171 CLR 506. See also Hart and Honoré (1985).

12   J 466.

However, she did qualify this position when it came to sentencing the defendants:

> "It is clear that the role of [the department] ... in providing the incorrect Record Tracings to the defendants and that [the department] has not been prosecuted has caused the defendants to feel 'a justifiable sense of injustice' ... Those feelings are understandable."[13]

She therefore accepted that the department's error was a mitigating factor in relation to the culpability of the defendants,[14] but she insisted that the companies were independently culpable for failing to verify the plans. That failure was essentially a failure to carry out an adequate risk assessment.[15] Inrush from old workings was a well-known hazard, with the potential to cause multiple fatalities. Had this risk been adequately assessed, the need for more effective risk control measures would have been obvious. The critical nature of the plans and the need to be certain of their accuracy would have been highlighted.[16]

In order to be clear about what was alleged against the companies, the prosecution charged them with a series of offences, but these were, for the most part, consequential on the initial failure to check the accuracy of the plans.[17] As the judge said:

> "[T]he genesis of many of the defendant's alleged failures derives from what I characterise as its primary failure to properly research the location and extent of the ... old workings. [T]o a large extent ... the particularised failures ... arise derivatively from the primary failure."[18]

The derivative nature of many of the charges was taken into account at the time of sentencing and, for present purposes, it is convenient to speak of a single offence.

The NSW OHS Act (under which the prosecution took place) also states that, where a corporation commits an offence, all those concerned in the management of the corporation are guilty of the same offence, unless they show that they were not in a position to influence the conduct of the corporation or, if they were, they used "all due diligence" to prevent the contravention. The court convicted both managers under this provision, as well as the second surveyor (the first surveyor

---

13 S 54 ("S" refers to the sentence handed down in *McMartin v Newcastle Wallsend Coal Company Pty Ltd & Ors* [2005] NSWIRComm 31 (11 March 2005). Numbers following an "S" refer to paragraphs). The reference to a justifiable sense of injustice is from *Nesmat v WorkCover Authority of NSW* [1998] 87 IR 312 (S 47, 48).

14 S 46, 48.

15 J 548ff.

16 There is considerable doubt as to whether a risk assessment would really have raised the question of the adequacy of the plans (I p 418).

17 See J 370 for a partial listing; third set of particulars at J 362.

18 J 633.

having died of cancer a month prior to the accident).[19] Conviction meant that these three individuals were found:

- to be concerned in the management of the corporation
- to have been in a position to influence the conduct of the corporation in relevant respects, and
- to have failed to exercise due diligence to prevent the contravention.

The first two of these points are not at issue here; what is of particular interest is the finding that the managers and the surveyor had failed to exercise due diligence, or in less formal language, appropriate care.

What was it that the judge expected them to have done? To answer this question, we need to know more about the nature of the error in the plans that had been given to mine management. The following description is rather detailed and the reader will need to grasp these details in order to comprehend later discussions.

Nature of the mistake

The old mine, worked until 1912,[20] was known as the Young Wallsend Colliery. Mining had taken place in two parallel coal seams, a top seam and a bottom seam, separated by 18 metres of other material.[21]

The Department of Mineral Resources provided Gretley management with two plans, one headed Young Wallsend Coal Workings *Bottom* Seam, and the other headed Young Wallsend Coal Workings *Top* Seam.[22] The two plans were identified as RT (Record Tracing) 523, sheets 2 and 3, respectively. Gretley intended to mine the *top* seam. Sheet 3, therefore, was the one on which it relied. Both sheets carried notations indicating that they had been created from an earlier single tracing, sheet 1. In other words, the documents provided were not simply photocopies of an original held by the department; they were photocopies of plans that had been *derived* from the original. The original tracing was not provided to Gretley management and it was never requested by mine management, even though it was available in the department.[23]

Sheet 1, the original tracing, was the key to what went wrong. It had been drawn up initially in 1892 and contained various amendments to incorporate additional work that had been carried out prior to cessation of mining in 1912.[24] But there was one very peculiar feature of this plan: it indicated two sets of workings or tunnels, one in red and one in black. Moreover, these two sets overlapped each

---

19   J 944.

20   J 389.

21   J 395.

22   J 393; S 221[60].

23   J 382, 811.

24   J 386.

other. Crucially, there was no indication on the plan as to what distinguished the red from the black. As the judge said: "Any person looking at RT 523 Sheet 1 could not help but wonder as to the precise import of the red and black workings and their relationship to each other."[25]

In 1980 another mining company, BHP, asked the department for separate plans of the old workings in the top and bottom seams. It was in response to this request that sheets 2 and 3 were produced.[26] The unknown person who drew up sheets 2 and 3 resolved the uncertainty about the meaning of the black and the red by assuming that one colour referred to the top seam and one to the bottom seam and that these two sets of workings at different levels had been superimposed on a single sheet when sheet 1 had originally been created. They could therefore be separated into two sheets. In separating them out, the person concerned made the further assumption that the red workings referred to the bottom seam and the black to the top. To repeat, there were no annotations or other indications on sheet 1 that justified these decisions and, as the judge noted, "on any view, that was a big call to make".[27] It needs to be emphasised just how extraordinary these decisions were. As far as can be ascertained, they were no more than guesses. As it turned out, the guesses were wrong. Both the red and the black workings were in the same seam — the top seam, the seam that Gretley was later to mine.[28] So it was that the plan relied on by Gretley management failed to show a significant section of the workings that had occurred in the seam. The disaster occurred when Gretley miners broke through into these unmarked workings.

How should the surveyors have behaved?

Against this background, we can now be more explicit about how, in the judge's view, mine officials ought to have behaved. Her view was based in part on expert evidence about what could be expected of a competent surveyor. The essence of that expert evidence was as follows.[29]

The two plans handed over to Gretley contained various puzzles. For instance, the purported workings in the bottom seam would have been almost impossible to ventilate. Moreover, every surveyor knows that transcription is liable to create errors. These plans had not been certified by the department and the possibility of error was correspondingly greater. A competent surveyor would, therefore, have asked the department for sheet 1, on which sheets 2 and 3 were based. An examination of sheet 1 would have raised doubts about whether it was intended to represent two sets of workings in different seams superimposed on each other.

---

25 J 388.

26 J 389.

27 J 401.

28 J 46. In fact, there had been very little mining in the lower seam.

29 J 457, 462.

For a start, there were no notations on sheet 1 to indicate that this was the case. Furthermore, looking at sheet 1 more closely revealed that, where the red and black workings overlapped, they did so exactly. If they had really been in different seams, there is no way that they could have aligned so exactly and, in any case, there was no particular reason to expect top and bottom tunnels to align at all. These and other anomalies raised serious doubts about the decision to separate the two sets of workings as the department had done. It was evident from an examination of sheet 1 that sheets 2 and 3 could not be relied on and that other methods would need to be used to determine the exact location of the old workings.[30] In the judge's words:

> "... the defendant and those acting on its behalf failed to recognise the glaring inconsistencies that Sheets 2 and 3 presented ... They were not such inconsistencies that a competent surveyor should have failed to recognise them. The evidence ... is compelling as to the extent of the basic surveying principles ignored by those charged with the responsibility to check such matters and who seemingly embraced Sheets 2 and 3 without question."[31]

She added:

> "[A]ny competent Mine Surveyor would not only ask to look at and obtain a copy of Sheet 1, but having done so, would immediately be alerted to the anomalies and irregularities in Sheet 1 and question the basis of the decision made within [the department] to separate and depict the red and black workings in the way that was done."[32]

The surveyor's failure to take any additional steps to verify the plans was, according to the judge, "the essence of (his) culpability".[33]

How should the managers have behaved?

The managers had relied on their surveyors and the judge recognised that this was, to some degree, inevitable. The expert witness had said:

> "The Mine Manager has the ultimate responsibility but would rely on the expertise of the other professionals available to him. He is in effect the captain of the team and he should not be expected to be a Surveyor, a Mechanical Engineer and an Electrical Engineer and so on."[34]

---

30   The additional research required was extensively discussed in the judgment but need not detain us here.

31   J 472, 473.

32   J 464.

33   S 263.

34   S 157.

The judge acknowledged this, but she went on:

> "Where [Mr P] failed, in my view, is that in taking steps to satisfy himself as to the accuracy of Sheets 2 and 3 in properly identifying the location and extent of the [old colliery], he too easily accepted the assurances he was given by the respective Mine Surveyors. He took no independent steps beyond that assurance. It was, I believe, a one off but serious error of judgment. In that respect, he failed to take the essential and fundamental independent step required of him in accord with the provisions of s 37(2)(h) of the [Coal Mines Regulation Act] that as Mine Manager he:
>
>> 'take such steps as may be necessary to ensure that at all times the manager is in possession of all available information ... regarding disused excavations or workings in the vicinity of the mine.'
>
> That responsibility was his and his alone. At the very least he owed it to himself to direct his Mine Surveyors, and any other relevant staff, to proactively research every available source for all information regarding the ... old workings. The implications if Sheets 2 and 3 were wrong demanded that. That is not something [Mr P] could delegate. Nor is it something that required additional skills and knowledge ... That to me is the essence of [Mr P's] culpability."[35]

In relation to the other manager, there was evidence that he had actively questioned one of the mine surveyors about the accuracy of the plans[36] but, as the judge noted, he did not look at the plans himself to "ascertain whether in his view they were reliable. In this respect he relied entirely on [the surveyor]".[37] His culpability was essentially the same as Mr P's.[38]

The penalties

The penalties imposed on the various defendants were as follows:

Operating company	$730,000
Owning company	$730,000
First mine manager	$30,000
Second mine manager	$42,000
Mine surveyor	$30,000

35 S 180-182.

36 S 221[64].

37 S 220.

38 S 227.

In order to understand the significance of these penalties, they need to be seen in relation to the maximum penalties available under the legislation at the time of the offence. The defendants were each convicted of several offences[39] and the judge imposed sentences for each offence. But since, as explained earlier, most offences were consequential on a single initial failure, she then discounted the penalties in various ways.

In the case of the companies, the maximum penalty was $500,000 per offence and the notional penalty assessed for several of these offences was $300,000 per offence — in each case, more than half the maximum possible. After discounting and splitting the fine equally between the two companies, the fine for each company came to $730,000. In this way, the total fine per company came to more than the maximum possible per offence.

For the individual defendants, the judge proceeded a little differently. The maximum for an offence was $50,000.[40] She determined a notional fine for each offence. But, instead of discounting and then adding the discounted fines to determine a final figure, she reversed the process. She selected a total which she intended to impose on each individual and then apportioned this penalty between the various offences in such a way that they added to the predetermined total. This procedure suggests that the judge was treating the $50,000 maximum not just as the maximum fine per offence, but also as the maximum fine per offender. As a result, the total fine imposed on each individual did not exceed the maximum fine per offence, as it did in the case of the companies. It is noteworthy, however, that the total fine for each individual was in the upper half of the $50,000 penalty range.

The sentencing rationale: culpability

I turn now to the sentencing rationale articulated by the judge when imposing these fines. The primary consideration was the degree of culpability of the defendants. Culpability is a very basic idea, not in any way limited to legal contexts; it means, quite simply, blameworthiness or fault. The question of interest here is how the law made judgments about culpability in the Gretley case. Precisely because culpability is an idea that is broader than the law, we are entitled to appeal to this broader, if intuitive, understanding when evaluating the legal criteria. I shall do this at various points in the discussion below.

The starting point for determining culpability, according to the judge, was the "objective seriousness of offence".[41] There were two main considerations: first, the potential consequences of the failure to exercise due care; and second, the foreseeability of the consequences. I shall deal with these in turn.

---

39   J 145, 204-206, 241, 242, 271, 272.

40   S 203.

41   S 39, 40.

Occupational health and safety legislation does not prohibit death and injury; it imposes a duty on employers (among others) to maintain a safe workplace as far as reasonably practicable.[42] The point is that a workplace may be patently unsafe and an employer therefore liable to prosecution, even though no death or injury has occurred. The fact of death or injury is not strictly relevant. It is the *potential* for death or injury that matters. As legal authorities have said, "the gravity of the consequences of an accident, such as damage or injury, does not, in itself, dictate the seriousness of the offence or the amount of penalty".[43]

It has to be said that this legal principle does not seem to operate in practice with anything like this clarity. Many regulatory agencies state that their policy is to prosecute when it is in the "public interest" and that the actual occurrence of death or serious injury is one of the factors taken into account when determining the public interest. Thus, for instance, the major OHS regulator in NSW, WorkCover, states in its guidelines that "WorkCover tends to prosecute when a death has occurred, when there has been a serious injury, or when there has been a risk of fatal or serious injury".[44] The policy of the NSW mines inspectorate is that, "generally speaking, the more serious an offence and the more serious its actual or potential consequences, the more likely it will be that the public interest will lead to prosecution".[45] It is clear from these statements that the actual occurrence of death or injury is indeed relevant to the decision to prosecute. In practice, it is rare for prosecutions to occur in the absence of such harm. Imagine that Gretley management had discovered its mistake at the last moment and stopped mining just in time to avoid the disaster, and assume that the inspectorate had been aware of what had happened. It is almost inconceivable in these circumstances that the earlier failure to check the accuracy of the plans or to carry out an adequate risk assessment in relation to the danger from old workings would have resulted in prosecution.

This last conclusion is supported by a statement made by the NSW Chief Inspector of Mines at the initial inquiry into the Gretley disaster. The de facto policy of the Department of Mineral Resources, he said, was to consider prosecution only when there had been a death or serious injury stemming from the violation, or when companies were failing to comply despite repeated warnings.[46] This is a far cry from the legislative intent.

---

42 Although the situation in NSW is formally different, I shall argue later that the difference is not as significant as is sometimes suggested.

43 S 21.

44 NSW WorkCover, *Compliance policy and prosecution guidelines*, March 2004, section 5.9.

45 NSW Department of Mineral Resources, *The enforcement of health and safety standards in mines*, January 1999, section F.5.1.

46 I p 694. Even this de facto policy appeared not to have been followed. In the seven years from 1990, there had been 33 deaths in NSW coal mines but not a single prosecution (I p 686).

But to return to the judge's reasoning, regardless of whether anyone was actually killed, the potential consequences of the failure to check the accuracy of the plans were extremely serious. This increased the seriousness of the offence itself, that is, the level of culpability.

Forseeability was the second main factor taken into account by the judge when determining the seriousness of the offence. She quoted approvingly from another judgment, as follows:

> "[T]he degree of foreseeability is a significant factor to be taken into account when assessing the level of culpability of the defendant. The existence of a reasonably foreseeable risk to safety which is likely to result in serious injury or death is a factor which will be relevant to the assessment of the gravity of the offence."[47]

The defence argued that the judge needed to make a distinction between the foreseeability of inrush in the vicinity of old workings and the foreseeability that the plans were incorrect, and that it was the forseeability that the plans were incorrect which needed to be considered when determining culpability. Inrush was a risk that was foreseeable and indeed foreseen, and mine officials had taken steps to control this risk by mining in accordance with the plans that they had been given. What was not foreseen was an error in the plans. This was rather less foreseeable than the risk of inrush itself.[48] The culpability was correspondingly less.

The judge rejected this reasoning. The question was "whether there was 'an obvious or foreseeable risk to safety against which appropriate measures were not taken'".[49] The answer was, yes, inrush was an obvious risk, yet the measures taken were far from adequate. In the circumstances of such an obvious risk, the mine should not have been relying unquestioningly on the plans provided.

It has to be said that this is a curious piece of reasoning. The judge had determined that the culpability of the surveyor lay in his not being alert to the possibility that the plans might be in error. A competent surveyor would have foreseen that possibility and taken steps to verify the accuracy of the plans, she said. The surveyor's culpability lay in his failure to foresee a reasonably foreseeable error. Similarly, according to the judge, the culpability of the managers lay in failing to seek independent verification of the accuracy of the plans. The desirability of seeking independent verification logically entails foresight of the possibility that the plans might be in error. If error in the plans is not a foreseeable possibility, why bother to verify them? I shall assume in the

---

47 S 19.

48 S 42.

49 S 44.

following discussion that the relevant failure of foresight was the failure to foresee that the plans might be in error.[50]

Another factor that is relevant to culpability is whether or not the offence is part of a general pattern of failure to attend to risk or, on the contrary, an isolated aberration.[51] The judge found that:

- the defendants, both corporate and personal, were generally safety conscious

- the operating company had an effective safety management system

- there was "an active workplace safety culture among employees and corporate defendants"

- workers were encouraged to cease work when they encountered a hazard,[52] and

- there was a "genuine commitment to workplace safety" on the part of all defendants.

Conversely, "there is no evidence that suggests the defendants had overall demonstrably unsafe systems of work or were indifferent as to their workplace responsibilities in relation to safety".[53] This led the judge to conclude that the failure to properly research the location of the old workings was "a most significant and serious lapse in an otherwise demonstrable commitment to providing a safe workplace".[54]

This last conclusion is clearly overstated. A good safety management system is no guarantee against lapses and errors of various sorts. Every thorough safety system audit finds problems; an audit which doesn't is hardly credible (Hopkins, 2000:80). It is doubtful that the lapse identified in this case was unique and no sensible manager would ever make such a claim. The comment must be seen as the judge's way of giving credit where credit was due and treating the

---

50   Bluff and Johnstone (2005:201) summarise the general position as follows: "Proving that the duty owed was breached requires the court to determine, on an objective basis, first, whether the risk was one that the defendant should have considered taking measures to guard against; and second, the measures that a reasonable person in the position of the defendant should have taken to control the risk." The question, then, is what measures the reasonable person would have taken to control a risk. Suppose a person reasonably assumes that certain precautions will be adequate but, due to circumstances that are not reasonably foreseeable, the precautions turn out to be inadequate. In this case, the person has done what the reasonable person would have done and there can be no breach of the duty. Putting the matter thus reveals that foreseeability enters twice into the determination of liability. There is, first, the foreseeability of the initial risk, and second, the foreseeability that the measures taken to control this initial risk will prove inadequate. It is arguable that, if foreseeability is relevant in two ways when determining liability, it is relevant in these same two ways for the purposes of assessing culpability at the time of sentencing.

51   S 62.

52   S 62, 63.

53   S 64.

54   S 85.

company's generally laudable approach to safety as a mitigating factor, thus reducing the level of culpability.

The other mitigating factor (already mentioned) was the role of the Department of Mineral Resources in providing incorrect information. In a sense, the judge had found that the department should share some of the blame for what had happened. But despite these mitigating factors, she still assessed the culpability of all of the defendants as being "towards the high end of the range of penalty available".[55]

The sentencing rationale: retribution

The concept of retribution was discussed in an earlier chapter and was equated with the idea that punishment is required so that offenders receive their just deserts. Furthermore, the need for retribution was seen to be directly proportional to the degree of culpability.

However, at one point in her sentence, the judge in the Gretley case made a distinction between culpability and retribution. She stated that the penalty must properly reflect the defendant's "culpability and society's retribution".[56] In saying this, she appeared to be suggesting that penalty would be determined not only by the level of culpability, but also by society's need for retribution. She did not expand on this second purpose, but it invites consideration.

The NSW *Crimes (Sentencing Procedure) Act 1999* (which guided the judge in the sentencing process[57]) contains a statement about the purposes of sentencing.[58] Retribution is not explicitly included among these purposes. The statement does, however, include the following two purposes: (1) to ensure that the offender is adequately punished for the offence; and (2) to recognise the harm done to the victim of the crime and the community.

Arguably, the first of these refers to the defendant's culpability and the second to society's need to see the offender punished. If so, they provide a basis for the judge's distinction.

There is, of course, a problem with this distinction. It suggests that society's need to see offenders punished is to be given weight independently of the question of culpability or, at least, over and above the question of culpability. This is hardly tenable. No court could responsibly impose a penalty that was higher than was warranted by the degree of culpability simply because of societal demand. Nevertheless, in the Gretley case, there was public pressure to see significant punishments imposed. Realistically, there was only one way in which the court could impose substantial punishments — and that was if the defendants were

---

55   S 88.

56   S 199.

57   S 16.

58   Section 3A. See discussion in ch 2.

found to be highly culpable. There is no way of knowing whether the need to satisfy public opinion influenced the judgments about culpability in this way. All we can say is that the fact that the defendants were in fact found to be highly culpable meant that society's need for retribution could be satisfied.

The sentencing rationale: remorse

Earlier discussion referred to the *objective* seriousness of the offence as the basis for judgments about culpability. The judge noted that there were additional *subjective* features which impacted on culpability. Remorse and contrition were singled out for particular consideration. These two ideas are similar; arguably, they are indistinguishable. The judge did not distinguish between the two and I shall not do so here.

In principle, remorse can affect a sentencing decision in two ways (Hall, 1994:B221). First, it may reduce the gravity of the offence in the eyes of a judge, so that purposes of retribution may be served by a lesser sentence. Second, if the offender is truly remorseful, it means that some measure of rehabilitation has occurred, that further offences are less likely, and that the need for specific deterrence is therefore lessened. Human beings are capable of remorse, but it is hard to conceive of a remorseful company. Nevertheless, companies were in the dock in the Gretley case and the judge was obliged to give some attention to the concept of corporate remorse.

When Esso was prosecuted in 2001, following the gas plant explosion at Longford in 1998, the judge found that Esso had failed to exhibit any remorse (Hopkins, 2002). He took this view on three grounds. First, the company had engaged in litigious treatment of its employees, blaming them for the explosion. Second, the legal defence it had advanced was one of obfuscation, designed not to clarify but to obscure. Third, the company had failed to accept responsibility for the explosion.

These conclusions were put to the judge as being relevant in the Gretley case.[59] She found, however, that the Gretley companies had not engaged in litigious treatment of their employees and that their defence, while vigorous, was not designed to obscure. On the question of whether the companies accepted responsibility, the evidence was not as stark as it had been in the Esso case.[60] Nevertheless, the defendants in the Gretley case clearly felt that the department was to blame: "In that sense the defendants, both corporate and personal (with the exception of [Mr P]) [of which more below], have demonstrated a reluctance to accept full and practical responsibility."[61] Of course, they deeply regretted what had happened but, in the absence of any acceptance of responsibility, this

---

59   S 133-143.

60   S 140.

61   S 141.

did not amount to remorse.[62] Accordingly (apart from Mr P), there were no grounds here for mitigating the penalties.

As for Mr P, the manager at the time of the accident, under questioning, he explicitly accepted responsibility and expressed remorse. He spoke about the searing impact which the accident had had, and continued to have, on his life. The judge stated that this was "the fullest expression of remorse and contrition [she had] ever encountered in circumstances of a workplace accident".[63]

It is difficult to gauge the impact of the remorse factor on penalties. The judge was at pains to stress that the absence of remorse did not *increase* the penalties.[64] But to what extent did Mr P's expression of remorse *reduce* the penalty imposed on him? In fact, he was fined $42,000, while the other manager and the surveyor were each fined $30,000. The differing fines reflect the differing circumstances in which these individuals found themselves; Mr P was the manager at the time of the accident. But one thing is clear: Mr P's expression of remorse did not save him from the judge's conclusion that he was the most culpable of the three.

The sentencing rationale: deterrence

Although the primary factor determining the penalty was the seriousness of the offence, it was not the only factor to be taken into account. Punishment was also expected to have a deterrent effect, both on the offenders concerned (specific deterrence) and on potential offenders (general deterrence).[65] The need for such deterrent effects influences the penalty: where the need for deterrence is considerable, the penalty can be increased; where it is minimal, the penalty can be reduced. Let us consider how the issue of deterrence affected sentencing for corporate and personal defendants, in turn.

In relation to the companies, the judge accepted that the industry as a whole had learnt the lessons concerning the need to be particularly vigilant about the risk of inrush, and that companies now approached old mining plans with the utmost caution. From this point of view, the general deterrent effect which punishment might be expected to have on the industry as a whole had already been achieved. However, in the judge's view, this did not totally eliminate the need for general deterrence. She quoted approvingly the following words from another judgment:

> "[T]he fundamental duty of the Court in this important area of public concern ... [is] to ensure a level of penalty for a breach as will compel attention to occupational health and safety issues so that persons are not exposed to risks to their health and safety at the workplace."[66]

---

62   S 234-236, 268.

63   S 191.

64   S 141, 236.

65   S 22.

66   S 122.

The reasoning here seems paradoxical: if the general deterrent effect has already been achieved, there is no need for further general deterrence. However, if the aim is to make companies more careful about hazards in general, not just the hazard of inrush, then the court's reasoning remains coherent.

As for specific deterrence, some time after the accident, the defendant companies were acquired by the mining multinational, Xstrata. Although this company had nothing to do with the accident, the judge took the view that specific deterrence was relevant to Xstrata:

> "The evidence . . . as to Xstrata's comprehensive system of workplace safety standards, while commendable, highlights factors that, in my view, reinforce the need for specific deterrence to be taken into account in the current sentencing process . . . The defendants are part of a large corporate enterprise that must maintain constant vigilance and take all practicable precautions to ensure safety in the workplace . . . While steps taken by the defendants to date post the inrush are to be commended, the scope of the defendant's ongoing obligations requires the need to encourage a sufficient level of diligence by the defendants in the future."[67]

Since Xstrata was in no way responsible for the accident, these comments about specific deterrence are a little perplexing. It is arguable that Xstrata occupies the same position as any other mining company in this matter and that, in so far as the penalty can be expected to have any deterrent effect on Xstrata, it is better viewed as a general rather than a specific deterrent effect.

The conceptual difficulties highlighted in the previous discussion would be of concern if it was clear that considerations of deterrence were significant in the final decision on penalties for the corporate defendants. There is no evidence, however, that they were. As noted earlier, in the judge's view, the offence in and of itself merited a penalty towards the top end of the range and that is what they received.

The situation in relation to the individual defendants was a little different. Overall, the judge accepted that none of them was as culpable as the corporate defendants, since each had played only a part in determining the outcome.

As for deterrence, her view was that "general deterrence had some, albeit limited, application".[68] Presumably, what she meant was that the sentences were intended to send a signal to other mining officials of the need to be diligent with respect to safety. She was clear, however, that there was no need for the sentence to have any specific deterrent effect on the individuals concerned. All three had clearly learnt the lesson and needed no further incentive. All had suffered greatly. Moreover, all had undergone various "rehabilitative measures". The two mine managers had completed risk management courses and the surveyor had

---

67   S 131.

68   S 184, 229, 264.

attended various lectures and retraining seminars for surveyors.[69] Despite this, the individual defendants received fines in the upper half of the penalty range. It would seem, then, that the issue of deterrence played a relatively small part when it came to the determination of sentence for the individual defendants.

Culpability reconsidered

Having described the considerations which went to the sentencing of the corporate and individual defendants, let us return to the question of culpability and consider, in a little more detail, what it meant in the Gretley case.

Judgments about culpability are judgments about moral rectitude, about the degree of wickedness, if you like. These are fundamentally human characteristics. It is very difficult to talk meaningfully about the moral rectitude or wickedness of a corporation.[70] There is a famous saying that a corporation "has no body to be kicked or soul to be damned". We can talk about whether a corporation is law abiding, whether it is deterrable, and whether, from a policy point of view, it is useful to hold corporations liable and to punish them.[71] But it makes little sense to talk about whether a corporation, as such, is blameworthy (Cressey, 1989; Braithwaite and Fisse, 1990). There are various solutions to this problem. The law recognises that "a corporate employer can only conduct its activities through human agents, such as its managers, supervisors, employees and contractors" (Johnstone, 2004:229). Thus, to say that a corporation failed to foresee what was reasonably foreseeable is to say that relevant managers failed to foresee what was reasonably foreseeable.[72]

Transferring the culpability of individuals to corporations in this way seems artificial, indeed, counter-intuitive. But the criminal law generally requires a finding of culpability before heavy punishment can be imposed, and transferring individual culpability to the corporation is one way that this can be achieved. In this sense, corporate culpability is simply a derivative of individual culpability.

What, then, can we say about individual culpability in the Gretley case? Note, first, that judgments about culpability are not factual judgments. They cannot be made by reference to evidence alone. A judgment about culpability is a value judgment, about what people ought or ought not to have done. Implicit in such judgments is a comparison with how some other reference individual would

---

69   S 187, 230, 265.

70   Bakan (2004) makes the argument that corporations are inherently criminal in the sense that they have no commitment to law-abiding behaviour and remain law abiding only in so far as it is in their interest to do so.

71   There is, of course, a huge body of literature on this. As a starting point, see Fisse and Braithwaite (1993).

72   Where a corporate offence requires proof of a guilty subjective state of mind (eg intent), only the state of mind of someone who can fairly be described as the directing mind and will of the corporation can be taken as the state of mind of the corporation. See Johnstone (2004:206).

have behaved. To blame someone is to say that their behaviour in some way falls short of the behaviour of the reference individual.

Let us explore in a little more detail the characteristics of this implied reference individual. Recall that the judge talked about a *competent* surveyor and concluded that the surveyors at Gretley had not behaved as a competent surveyor would have. The inference is that they were incompetent. But incompetence is hardly a moral evaluation. If they were indeed incompetent, they can hardly be blamed for this. The fault lies with the education system which had certified them competent, or with a management that had not verified their competence for the job at hand. The fact that the behaviour of the surveyors fell short of how a competent surveyor would have behaved hardly makes them culpable. In short, the *competent* surveyor is not the reference individual we are looking for.

In finding the individual defendants culpable, the judge was explicitly finding that they had failed to exercise "all due diligence". How would a person who was exercising all due diligence be behaving? According to Johnstone:

> "... due diligence is generally accepted as the converse of negligence and requires reasonable care in all the circumstances ... What is due diligence will depend on all the circumstances of the case, and contemplates a mind concentrated on the facts. It is to be decided objectively according to the standard of the reasonable person in the circumstances" (Johnstone, 2004:210-211, citations omitted).

Here, then, is our reference individual — the reasonable person in the circumstances, the reasonable surveyor or the reasonable manager.

What can we say about these reference individuals contemplated by the court — the reasonable surveyor and the reasonable manager? Are they, for instance, the ordinary[73] or average[74] surveyor and manager? What we know is that neither of the two managers and neither of the two surveyors behaved in the way that the judge believed the reasonable person in the circumstances would have behaved. All four accepted the accuracy of the plans provided. Were all four unreasonable men?

The Gretley defendants accepted that old plans might have small errors in them, which could arise in the following way. Mines which are being actively worked must update their mine plans every three months.[75] If mining at an old colliery was terminated between updates, some workings would have remained

---

73 The ordinary person test is used in the law of provocation and has been subjected to much criticism in this context. See Bronitt and McSherry (2005:273-281). There is no need to dwell on this issue in the present context.

74 The concept of an average human being is problematic. What is the sex of the average human being? Strictly speaking, the "mode" (meaning the "most common") would be a better term here.

75 I p 231.

unrecorded. Errors of up to 26 metres had been known to occur in this way.[76] To allow for this, Gretley mine management intended to leave a barrier of 50 metres between the current workings and where the old workings were thought to be. Apart from errors of this nature, the assumption was that the plans provided by the Department of Mineral Resources were accurate and could be relied on.

The evidence suggests that other mine managers and surveyors had made similar assumptions. The Gretley mine manager who immediately preceded the two who were charged said that he had accepted the accuracy of the plans provided by the department: "That was always our position and I believe the view of the mining industry, up until the inrush itself."[77] An earlier surveyor at Gretley had accepted the accuracy of the plans and had certified them as accurate.[78] The chief company surveyor also assumed that the plans were accurate.[79]

Two mines adjacent to Gretley had made the same assumptions.[80] They had been provided with the same departmental plans which their surveyors had duly certified. This was known to Gretley managers, which reinforced their assumption that the plans were accurate.

The NSW Chief Inspector of Mines had also assumed the accuracy of the plans: "Up until this accident, I regarded Record Tracings supplied by the department as accurate representations of old workings. I had never previously seen a Record Tracing that I knew to be inaccurate."[81] Finally, a senior inspector gave evidence that he had assumed that the plans provided by the department were accurate[82] and he noted that this assumption was "well entrenched and accepted — both at Gretley and adjacent mines".[83]

An earlier inquiry conceded all this. It accepted "that a sizeable number of individuals within the mining industry assumed before the inrush that 50 metres . . . offered adequate protection against inadequate plans".[84]

It is clear that many, perhaps most, surveyors and managers would have behaved as the defendants did. If that is so, the ordinary mine surveyor and manager does not qualify as reasonable in the judge's view; the reasonable officials she had in mind are probably few and far between.

---

76   I p 45.

77   I p 241.

78   I p 271.

79   I p 243.

80   S 173; I p 249.

81   S 58[19].

82   I p 50.

83   I p 419.

84   I p 243.

This is a significant departure from community perceptions of reasonableness (Luntz and Hambly, 2002:244). The reasonable person was once famously defined as the man on the Clapham omnibus and, more recently, as the traveller on the London Underground.[85] Such descriptions are intended to convey that the reasonable person is an ordinary person, not an unusually diligent or careful person. Yet this is precisely what the judge was requiring of the individuals before her — that they be unusually diligent and careful, indeed, that they behave as the *ideal* surveyor or manager would behave.[86] If the reference individuals in the judge's mind were the ideal surveyor and the ideal manager, it is to be expected that real-life surveyors and managers will fall short of these ideals. To conclude on this basis (that real-life individuals have behaved unreasonably and therefore culpably) seems unfair.

Reason (1997:208) has developed this line of thinking into a test for whether or not it is appropriate to blame an employee who violates rules. He terms it the "substitution" test. Mentally substitute the individual concerned with someone else who has the same training and experience and ask: "In the light of how events unfolded and were perceived by those involved . . . is it likely that this new individual would have behaved any differently?" If the answer is no, there are clearly systemic factors generating the behaviour concerned and it is better to seek to change those factors than to blame the individual rule violator. This principle applies to managers as much as it does to front-line workers, and it provides a rationale for rejecting the culpability of the individual Gretley defendants for relying on faulty plans.

For all of these reasons, it is hard to agree with the judge that the culpability of the individual defendants for relying on faulty plans is so serious as to put them "towards the high end of the range of penalty available".

Two final points need to be made. First, none of this is to suggest that the behaviour of those concerned was good enough. With hindsight, it is clear that the standard of care being applied was quite inadequate. The industry has learnt this lesson well and NSW mines are now subject to stringent requirements that are aimed at avoiding the possibility of inrush. The bar has been raised. Were there to be any repeat of the Gretley disaster, those concerned would be unequivocally culpable.

Second, this analysis does not imply that it was inappropriate to hold individuals accountable for the death of the Gretley miners. Chapter 8 will present an extended argument on the desirability of personal accountability in such circumstances and how this might be achieved. The point here is simply that, on the basis of the evidence presented so far, the conclusion that these individuals are blameworthy and therefore deserving of punishment seems somewhat unfair.

---

85   *Cattanach v Melchior* [2003] HCA 38 (16 July 2003), para 82, 135.

86   The law frequently departs in this way from the idea that the reasonable person is the ordinary person. This will be discussed in greater detail in a later chapter.

Chapter 4

The second failure: dismissing the warnings

As we saw in the preceding chapter, courts seek to make judgments about culpability. Where an employer has failed to take effective steps to manage a risk with potentially serious consequences, then the more foreseeable the risk, the more culpable the defendant. In short, given a potentially serious risk, forseeability is the crucial test of culpability. I have shown that the judge in the Gretley case focused on the question of whether or not it was foreseeable at the *outset* that the plans provided by the Department of Mineral Resources might be in error.

The whole situation changed, however, in the weeks immediately prior to the inrush. Indications began to emerge that the Gretley miners were closer to the old workings than the plans indicated. These indications were considered but effectively dismissed. Had they been taken more seriously, the accident could have been averted. The question to be addressed here is whether these indicators made it reasonably foreseeable in the last few days that the plans might be in error.

The Gretley disaster generated two sets of findings. In 1997/98, an inquiry was held before Judge James Staunton. Its purpose was to establish the causes of the accident and it was not a forum for determining liability. One of its recommendations was that consideration be given to prosecuting the companies. The prosecution took place in 2003/04, before Judge Patricia Staunton (no relation to James). So far, the focus of this book has been on the findings arising from the prosecution. The present chapter draws heavily on the findings of the earlier inquiry, which will be referred to here as the "inquiry". Any references to the "court" are to the later prosecution.

As mining approached the old workings, various people reported that water was seeping out of the mine face and accumulating at a low point on the tunnel floor. The question is: how should these reports have been interpreted? One influential view is that "any water inflow in the vicinity of abandoned mines . . . should be considered a danger signal".[1]

However, matters were not quite so simple. Gretley was a wet mine, and flows of water out of the seam were to be expected. Nevertheless, there was something unusual about the flow of water. As one witness observed:

---

1   I p 515.

"Gretley is a very wet mine, all the districts are wet, the water percolates from one district to another because of the way it is developed. The funny thing about [the district where the inrush occurred] and I always commented when I went in there, it is . . . (a) pleasure to come into this district, it is the driest district in the pit. It was probably the only district in the pit without a pump."[2]

The appearance of water did not demonstrate conclusively that mining was close to the old workings because an increase of water flowing from the face could be expected as the old workings were approached, even though those workings were still at a considerable distance.[3] But, clearly, the reports were an indication of *possible* danger. Before proceeding further, it will be useful to list these reports.

The reports of water

1 November

Day shift safety officer, Mr M, made the following comment in his end of shift report: "Nuisance accumulation of water."[4]

4 November

Mr M made another, similar comment in his end of shift report: "Large amount of nuisance water."[5]

Two other shift safety officers told the inquiry that it was unusual for safety officers to make such comments in their end of shift reports and the inquiry came to the same conclusion itself, having perused a large number of end of shift reports.[6] Mr M explained later that, although he had used the term "nuisance" water, he was intending to draw management's attention to a potential danger.[7]

Also on 4 November, Mr M reported the accumulation of water to the mine manager who happened to be underground doing an inspection. He said to him: "There is water gathered in 7 cut-through. We are not close to the old mine are we?"[8]

That night, the night shift safety officer, Mr B, commented to the shift manager about the accumulation of water. He observed that the accumulation of water had increased since 1 November. He also told the shift manager's superior, the undermanager-in-charge, about the water and the fact that it appeared to be unusual. He said later: "It seemed to me that, for a panel [area of the mine] in

---

2   I p 514.

3   I p 514.

4   I p 523.

5   I p 534.

6   I pp 536, 537, 638, 639.

7   I p 619.

8   I p 528.

such good condition, there was more water laying around the floor than I would have expected."[9]

13 November

The day shift safety officer, Mr M, spoke about the water on various occasions to the driver of the mining machine, as well as speaking to two other miners. He asked them repeatedly: "Where's all that water coming from?"[10] In his end of shift report, he stated: "Coal seam giving out a considerable amount of water."

It was the next day, 14 November, that miners broke through into old workings.

Management response to the reports

Following the reports on 4 November, the shift manager examined the water and concluded that it might have been coming out of the floor.[11] His superior, the undermanager-in-charge, concluded that the increased quantity of water was to be expected as they were heading towards the old workings.[12] *His* superior, the mine manager, observed the water accumulation but told the safety officer that the old mine was about 200 metres away. The inquiry made the following comment on this response:

> "The misgivings of an experienced deputy about a serious potential hazard, namely inrush, ought to have made [the manager] pause, and reflect upon what was being said. Instead, he brushed Mr M's concern to one side, glibly referring to the plan. A warning went unheeded which, had it been taken seriously and investigated, may have exposed the [error in the plan]."[13]

Following the 13 November report, the shift manager immediately turned to the mine plan and measured what he thought to be the distance between the face and the old workings.[14] He discussed the matter with the safety officer who had made the report and decided that nothing further needed to be done. Precisely what the safety officer said that led to this outcome is in doubt. Both the safety officer and the shift manager subsequently told the inquiry that the safety officer had effectively contradicted his written report and said to the shift manager that the "considerable" amount of water was in fact only a "trickle". It was this that led the shift manager to do nothing.

---

9   I p 524.

10   I pp 600, 601.

11   I p 525.

12   I p 527.

13   I p 532.

14   I p 617.

The inquiry did not accept this version of the conversation and expressed the opinion that it was intended to protect the company.[15] The inquiry's view was that, whatever the safety officer may have said to the shift manager, he remained of the view that what he saw was a "considerable" amount of water — enough to give rise to concern: "He saw the link, or possible link, between the water and the old workings, and recognised that it may be a symptom of danger. He was right to do so."[16]

The inquiry went on to say that the shift manager's "investigation of the observations by Mr M were superficial. Having recognised from Mr M's report the symptoms of danger, they were dismissed too readily".[17]

To summarise, the inquiry characterised management's response to the warnings as "superficial", "glib", and certainly inadequate.

Differences between the inquiry and the court on the investigation of warnings

Let us return to the prosecution. How did the court deal with these matters? One of the charges laid against the companies was:

> "(h) a failure to investigate, adequately or at all, the [safety officers'] written reports on 1 November, 4 November and 13 November 1996 and two oral reports on 4 November 1996."[18]

The court inclined to the view that:

> "... the defendant did adequately investigate the reports of water ... Ultimately, it has to be said the steps and investigations undertaken to investigate the reports of water ... were made against the background of the presumed location of the ... old workings ... being 100 metres or more away."[19]

In other words, given the presumption that the plans were accurate, management acted reasonably in dismissing the warning signs. More cautiously, the court went on to say: "Overall, I am not satisfied that failure (h) has been established to the requisite standard."

The contrast between the judgments of the court and the inquiry is perplexing. Could it be an outcome of the different nature of the proceedings — one a criminal prosecution and the other a judicial inquiry? In a criminal matter, the prosecution is required to establish its case beyond reasonable doubt; that is the "requisite standard". The findings of the judicial inquiry, on the other hand,

---

15   I p 613.

16   I p 617.

17   I p 627.

18   J 632.

19   J 703, 704.

were made on the balance of probabilities. It was thus, in theory, easier for the inquiry than the court to draw adverse conclusions from the same evidence. However, the comments of the court suggest that this was not the real issue. Even on the balance of probabilities, the court would probably have dismissed the charge.

Another possibility is that the nature of the evidence put before these two tribunals was different. The comments made by the court suggest that the evidence available to it may have been more limited. A third possibility is that the difference is simply a demonstration of the somewhat uncertain nature of judgments about culpability: the court taking the view that the various managers at Gretley behaved reasonably in relation to the reports of water and the inquiry believing that they did not.

The decision to drill ahead

There is one way in which mine management could have been sure that the old workings were not closer than indicated on the plan — and that was to check what lay ahead by drilling a small diameter hole forward from where mining was taking place. Drilling could also have determined whether there were any unusual geological formations ahead.

The possibility of drilling ahead was discussed in the first week of November and the undermanager-in-charge made the final decision to drill ahead just before he went on leave on 8 November.[20] On 13 November, the day before the inrush, a management meeting ratified this decision. The minutes state: "Advance drill 60 m to prove ground to be driven."[21]

Why was the decision made at this time, just days after the first reports of water had been made? The undermanager-in-charge told the inquiry that "water played no part in the decision to drill".[22] But in other statements he made it clear that water was the precipitating factor. He recounted a conversation that he had had with a subordinate shift manager, to the best of his recollection at the end of October or early November, in which he had discussed the water that was building up:

> Shift manager: "Have you thought about putting a borehole in advance of the workings?"

> Undermanager-in-charge: "I hadn't planned to. We are still a long way from the old workings. It would be useful to put the hole in to prove the ground in front of us."[23]

---

20   I p 543.

21   I p 544.

22   I p 544.

23   I p 545.

Elsewhere, he spoke of drilling ahead as "giving us the added safety factor as to the presence of the old workings"[24] and being undertaken so that the mine did not "run into any surprises".[25]

One interchange at the inquiry was the following:

> Q: "Is not [the possibility that the water was coming from the old workings] the thing that provoked the thought that we had better just prove the ground ahead?"
>
> A: "It was part of it, I suppose, yes."[26]

Perhaps the most telling interchange was this:

> Q: "[The decision to drill ahead] was a step that you had suggested, amongst other reasons, as an insurance against risk?"
>
> A: "That's correct."
>
> Q: "The risk that the old workings might be closer than the plans in fact indicated, is that right?"
>
> A: "You could assume that, yes."
>
> Q: "And even though that risk to you may not have been a high risk at the time, it was perceived by you as risk, is that not right?"
>
> A: "As a minimum risk."[27]

The inquiry concluded, in relation to the undermanager-in-charge, that "the presence of water had brought to his mind the possibility, although he thought it minimal, that the old workings were closer than the plan indicated"[28] and, by implication, that the plan was wrong. The inquiry also made the following statement concerning the shift manager who had initiated the conversation above: "The [inquiry] believes that he did recognise the possibility that the plan may be inaccurate."[29] In short, according to the inquiry, both of these men foresaw the possibility that the plan might be in error.

Evidence was also provided to the inquiry that the miners themselves understood that the purpose of drilling was to check that they were not about to break through inadvertently into the old workings.[30]

The criminal court, on the other hand, did not draw the conclusion that anyone had foreseen the possibility that the plans might be in error. The judge stated that there was no evidence that allowed her to conclude with certainty that the

---

24   I p 546.
25   I pp 543, 554.
26   I p 548.
27   I p 549.
28   I p 554.
29   I p 557.
30   I pp 555, 556.

decision to drill ahead was connected to the reports of water.[31] She had before her the minutes recording the decision to "advance drill 60 m to prove ground to be driven". But, she said, no evidence had been provided about exactly what this meant. It seems, therefore, that in this case the difference between the two tribunals reflected differences in the evidence put before them.

The actions of the surveyor

In early November, the surveyor had asked someone where he might get independent evidence of the location of the old mine, and he was referred to an agency called the Mine Subsidence Board. The plans that were made available by the board turned out to be the very same plans that had been available at the Department of Mineral Resources, including sheet 1. But they were made available on microfiche and sheet 1, when printed out, was of such poor quality that it was discarded.[32] It will be recalled from chapter 3 that this was the sheet that showed both sets of workings on the same plan, which should have cast serious doubt on the accuracy of the plans on which mine management was relying. In the end, then, despite the surveyor's efforts to obtain independent verification of the location of the workings, his efforts were in vain.

Why did he go to this effort to check the accuracy of the mine plans? There was some dispute about the evidence but the inquiry made the following findings, "as a matter of probability". The surveyor had been present at the discussion between the two managers about the water and the need to drill ahead: "He recognised that drilling ahead was being suggested because there was the possibility that the plan may be inaccurate. He therefore decided to check the plan."[33] In short, the surveyor, like the undermanagers, foresaw the possibility that the plans were in error.

The court also commented on the surveyor's actions: "The inference arising from such an inquiry [to the Mine Subsidence Board] is that mine management, through [the surveyor], was responding to reports of water ... that were surfacing at about that time and that [the surveyor] was double checking the location of the ... old workings."[34] This is tantamount to inferring that mine management had foreseen the possibility that the plans might be in error. The court did not explicitly draw this inference. The failure to do so seems somewhat inexplicable.

---

31 J 702.

32 I p 592.

33 I p 587.

34 J 417.

Culpability

The inquiry was not a tribunal set up for the purpose of establishing culpability. However, its findings are directly relevant to this issue. Let us recall the earlier statement on which the court relied:

> "[T]he degree of foreseeability is a significant factor to be taken into account when assessing the level of culpability of the defendant. The existence of a reasonably foreseeable risk to safety which is likely to result in serious injury or death is a factor which will be relevant to the assessment of the gravity of the offence."[35]

In short, all other things being equal, the degree of culpability is directly related to the degree of foreseeability.

It will be remembered that, prior to the commencement of mining, no one foresaw the possibility that the plans might be in error and the question for the court was whether a reasonable person would have. That issue does not arise here. The inquiry found that, in the final days before the inrush, the people concerned actually foresaw the possibility that the plans were in error. There could hardly be better evidence that this was a reasonably foreseeable possibility. Based on the findings of the inquiry and the test of culpability spelt out by the court, it can be concluded that these individuals were indeed culpable for their failure to respond adequately to the warnings of danger that occurred in the days immediately preceding the inrush.

But that is not the end of the matter. We need to consider just what would have constituted an appropriate response to the warnings. Drilling ahead was one appropriate response. But the drilling was scheduled to commence a day or two after 14 November — the day of the inrush.[36] For drilling to have been effective, mining should have stopped until drilling had indeed proved the ground ahead.

Why didn't the managers concerned suspend mining? The answer appears to be that, although the undermanager-in-charge foresaw the possibility that the plans might be in error, he regarded this as a remote possibility. He therefore concluded that, even though the ground ahead had not been proved, the risk in continuing to mine was an acceptably low one. We are back again to the question of reasonableness. Is this a judgment which a reasonable manager would have made? If so, this would presumably diminish the culpability which arises from the fact that he foresaw the risk but allowed mining to continue.[37] If, however, we judge that a reasonable manager in the circumstances would have ordered work to stop immediately until the ground ahead had been proved, then the

---

35   S 19.

36   I p 556; J 701.

37   The question of what a reasonable person who has foreseen a risk would do is discussed in Brooks (1993:78ff). Strictly speaking, this applies to liability, not culpability, but the reasoning is relevant here. The matter is also addressed by Mason J in *Wyong Shire Council v Shirt* (1980) 146 CLR 40; see Bluff and Johnstone (2005:203).

manager's failure remains culpable. The court did not deal explicitly with this point, but the inquiry made the following comment: "[The undermanager-in-charge] gave no direction to suspend mining and monitor the build up of water, *as he ought to have done.*"[38] (emphasis added)

In so saying, the inquiry was clearly blaming this individual.

A second appropriate response to the warning signs would have been for the surveyor to thoroughly research the location of the old workings. The approach to the Mine Subsidence Board turned out to yield no new information. Did this absolve the surveyor of responsibility? The court did not pass judgment on this particular question, but the inquiry did:

> "He should not have stopped his investigation at that point. Once there was doubt in his mind, it was his duty, first to inform the manager, and secondly to resolve that doubt completely (or disclose to his superiors that it was incapable of resolution, because of the paucity of material)."[39]

The inquiry was critical of the surveyor for failing to check the plans adequately in the first place, but it regarded as especially blameworthy this failure to take effective action after the possibility was recognised that the plans might be in error: "It was a completely unacceptable risk for [the surveyor] not to sound a strong warning in early November (at the very latest) and recommend that development be suspended immediately pending proper inquiry and confirmation."[40]

It would seem, therefore, that as far as the inquiry was concerned, having foreseen the risk, the individuals were under an obligation to act effectively to counteract that risk. Their culpability lay in foreseeing the risk but failing to respond effectively.

At this point, it needs to be stressed that this discussion does not translate automatically into conclusions about liability under the OHS Act. The Act envisaged that, when a corporation commits an offence, individuals may be found guilty of the same offence — but only if they are "concerned in the management of the corporation". The undermanagers had been charged, along with the managers and the surveyor. But, after a long discussion of what it meant to be "concerned in the management of the corporation", the court found as follows: ". . . I cannot be satisfied beyond reasonable doubt that those personal defendants employed as Under Managers and Under Managers in charge at Gretley were persons concerned in the management of the corporations."[41]

---

38   I pp 557, 558.

39   I p 595.

40   I p 505.

41   J 934. Had they been charged under section 19(a) of the OHS Act, the outcome might have been different. That section states: "Every employee while at work — (a) shall take reasonable care for the health and safety of persons who are at his or her place of work and who may be affected by his or her acts or omissions at work."

All charges against them were therefore dismissed. No matter how culpable they might be, they were not liable under the Act. In contrast, the court found that the surveyor and the mine managers were persons "concerned in the management of the corporation" and went on to convict them and make judgments about their culpability.

Conclusion

The court was not critical of the failure to respond effectively to the last-minute warnings. It concluded that the investigations of the reports of water were adequate in the circumstances, and that the failure to drill ahead in time was irrelevant, since it had not been proved beyond reasonable doubt that the decision to drill ahead had anything to do with the reports of water. Nor was the failure to effectively investigate the accuracy of the plans at this stage especially culpable. These failures were all, in the court's view, a consequence of the initial failure to investigate the accuracy of the plans prior to the commencement of mining. It was this initial failure that the court found so culpable, despite the fact that the managers and surveyors had behaved as many other ordinary managers and surveyors would probably have behaved in the same circumstances.

However, the evidence presented to the inquiry led it to a different conclusion, namely, that the failure to respond to the warning signs was indeed culpable. The inquiry judge did not use the word "culpable" but that is the inevitable implication of his language. His view with respect to culpability was almost the reverse of the court's, in that he appeared to conclude that the initial failure to investigate the accuracy of the plans was not as culpable as the failure to take effective action once the warnings emerged. Whether or not this alternative view is accepted, it is clear that tribunals can differ greatly in their judgments about culpability.

The sense of injustice generated by the prosecution was because the defendants had been blamed for their failure to check the plans that the Department of Mineral Resources had given them. Had the court taken its lead from the inquiry and blamed the defendants for their failure to respond effectively to the warning signs in the days immediately preceding the disaster, one wonders if the sense of injustice would have been as intense.

Chapter 5

The campaign to change the law

Following the conviction of the Gretley defendants, the parent company, Xstrata, mounted a broad-brush appeal to the NSW Court of Appeal, arguing that the NSW OHS Act was unconstitutional and a radical departure from established legal rights and principles. Two other mining companies mounted similar appeals arising out of unrelated cases. These appeals were all ultimately dismissed or discontinued.[1] Xstrata also mounted a more limited appeal, concerning the specific details of the case, before the Full Bench of the Industrial Court of NSW. As noted earlier, the full bench upheld the convictions against the two corporate entities, as well as the conviction of the manager at the time of the disaster. It varied the findings in relation to the former manager, finding him guilty on fewer charges and, on this basis, it chose not to proceed to conviction. The court quashed the conviction of the surveyor on the grounds that he had no role in management.[2]

The convictions also triggered a far more widespread response. An organisation calling itself Employers First mounted a campaign specifically against the NSW legislation. It claimed that:

> "There is no doubt that employers/directors/managers who have failed to provide perfectly safe workplaces with zero risk, will be sent to gaol — after having been deemed guilty before they even go to court, denied the presumption of innocence and the right to a jury trial, suffered the reverse onus of proof and the suicidal job of trying to prove the impossible defences under the Act."[3]

The Institute of Public Affairs, which describes itself as Australia's leading free market think tank, also weighed into the fray, complaining that, under existing NSW law, managers "are presumed guilty until proved otherwise. Incredibly, they have to prove they are innocent".[4]

---

1 *Powercoal Pty Ltd & Peter Lamont Foster v Industrial Relations Commission of NSW & Rodney Dale Morrison* [2005] NSWCA 345; *Coal Operations Australia Ltd v Industrial Relations Commission of NSW & Rodney Dale Morrison* [2005] NSWCA 346; and *Newcastle Wallsend Coal Co Pty Ltd v Industrial Relations Commission of NSW & Anor* [2006] NSWCA 129 (30 May 2006).

2 *Newcastle Wallsend Coal Company Pty Ltd & Ors v Inspector McMartin* [2006] NSWIRComm 339 (5 December 2006). See ch 2, fn 38.

3 Website at www.employersfirst.org.au/MAIN/NEWS_campaign_Advise.htm. Accessed 23 August 2005.

4 Phillips, K, "Guilty until proved innocent", *Australian Financial Review*, 23 November 2004. Available on the Institute of Public Affairs website at www.ipa.org.au/files/news_851.html.

The Australian Chamber of Commerce and Industry (ACCI) released a major policy statement that was profoundly critical of OHS legislation in Australia, both in the way it was drafted and in the way it was being interpreted. The ACCI's statement did not refer specifically to Gretley, but the text emphasised the situation in the state of NSW. The statement was described as a blueprint for the next 10 years and among the reasons it gave for releasing its blueprint at this time was the following:

> "There is a lack of balance in some existing legislation and court decisions. The trend across jurisdictions has been to broaden legal duties beyond reasonable limits, increase penalties, extend liability to individuals in the management and supply chain and to seek to punish rather than prevent."[5]

The ACCI's policy statement was the most detailed criticism made in the campaign and this chapter will seek to evaluate some of its arguments. I address in particular the claims about:

- reasonableness
- the reverse onus of proof in criminal matters, and
- absolute liability.

The chapter concludes by examining government's response to this campaign.

Reasonableness

It will be recalled that the standard of reasonable foreseeability that was applied in the Gretley case was such that most ordinary mine managers would not qualify as reasonable. The ACCI took up this theme: "The concepts of 'reasonably practicable', 'foreseeable' and 'control' have been significantly distorted in several Australian jurisdictions, to the point where they no longer reflect what is reasonable, practicable and achievable."[6]

It continued: "To foresee the unforeseeable, to know the unknowable and to control the uncontrollable is simply not reasonable."

These statements were inspired by the situation in NSW — specifically by the findings in the Gretley prosecution. The ACCI is implicitly arguing that the Gretley judge was expecting the managers and surveyor to foresee the unforeseeable or, perhaps more precisely, to foresee what was not reasonably foreseeable.

It is important to note that this issue goes beyond the Gretley case — indeed beyond OHS law. The law of negligence has moved well away from the idea of the reasonable person as simply being an ordinary person. The point was made very dramatically in the High Court some decades ago:

---

5   Australian Chamber of Commerce and Industry (ACCI), *Modern workplace: safer workplace. An Australian industry blueprint for improving occupational health and safety 2005-2015*, April 2005, p 11.

6   Ibid, p 21.

"There has been a tendency in cases of this type to forget the legal standard of reasonable care, and to regard the standard employer as a person possessing super-human qualities of imagination and foresight ... it is wrong to take as the standard of comparison a person of infinite-resource-and-sagacity."[7]

According to another judge, the hypothetical reasonable person has been "attributed with the agility of an acrobat and the foresight of a Hebrew prophet" (Luntz and Hambly, 2002:244).

Most recently, in 2005, a High Court judge again bemoaned the state of Australian law in this area.[8] An earlier judgment by three judges of the High Court had "held that any risk, however remote or even extremely unlikely its realisation may be, that is not far-fetched or fanciful, is foreseeable". Reluctantly, the judge said that he was bound by this rule, and he went on:

"With enough imagination and pessimism it is possible to foresee that practically any misadventure, from mishap to catastrophe is just around the corner ... The line between a risk that is remote or extremely unlikely to be realised and one that is far-fetched and fanciful is very difficult to draw. The propounding of the rule relating to foreseeability in the terms [above] ... requires everyone to be a Jeremiah [the Judean prophet of doom]."

Given that there has been no shortage of judicial critics of the way in which the reasonable person test has been interpreted, the questions are: why has this happened, and why have the reasonable and the ordinary person diverged to such an extent?

One of the purposes of the law of negligence is to ensure that, when a person is harmed as a result of someone else's negligence, the victim receives compensation from the negligent party. Those who run the risk of being sued for damages in this way are often insured. Actions for damages can thus be seen as distributing the loss by shifting it to the insured party, and thence to all those in the insurance pool. According to one view, this has become the primary purpose of the law in this area.[9] The higher the standard of care that the courts require, the greater the chance that the defendant will be found negligent, and hence the greater the chance that the victim will receive compensation. There is therefore a tendency over time for courts to impose higher standards of care in order to achieve this outcome. Judges have acknowledged that this is what they are doing and have argued that this is "a convenient and desirable means of achieving this policy objective".[10] But, as Luntz and Hambly (2002:224) noted, when used as a

---

7   *Rae v The Broken Hill Pty Co Ltd* [1957] HCA 33; (1957) 97 CLR (3 June 1957).

8   *Koehler v Cerebos (Australia) Ltd* [2005] HCA 15 (6 April 2005), para 54.

9   Stepinak, D, "Minority values and the reasonable person of torts" (unpublished paper). See also Flemming (1984).

10   Ibid, p 2.

justification to achieve this end, the test becomes "an idealised ethical standard, rather than the behaviour of the actual 'man in the street'".

The litigious society

The loss-spreading function of negligence law is associated with what has been described as a "culture of litigation". The former Chief Justice of the High Court, Sir Harry Gibbs, has written critically about this culture and his views are worth quoting at some length:[11]

> "Until after the first half of the twentieth century, people were disposed to accept mischance as part of life ... There has been a steady development in the community of the belief that if someone suffers an injury someone else must be responsible and must be liable to pay compensation ... The intoxicated customer of a club who goes outside and is struck by a motor vehicle sues the club; the worker who is mugged while leaving his place of employment sues the employer; the person injured while playing sport sues the organiser of the game or the person who devised the rules; the youth who sustains injury by diving into shallow water sues the local council or the owner of the land from which he dived. The injured person does not put it down to bad luck, or blame himself, but blames someone who can be sued."

There is, he says, an "unhealthy culture of blame, with [an] emphasis on rights rather than responsibilities".

Gibbs goes on to deride a section of the legal fraternity which has exploited this situation:

> "The culture of litigation has been fostered by some lawyers who, whenever a mishap occurs, advertise for potential claimants and encourage them to sue or to join in a class action. These activities would once have been regarded as unprofessional ... It would, however, be too much to hope that any ban on advertising or touting for work by lawyers would change the culture of blame which is now well established in society. The lawyers jumped on the bandwagon; they did not start it."

Gibbs' comments suggest that the problem is one of moral decline on the part of the citizenry — a refusal to accept responsibility for our own part in misfortune that befalls us, even an unwillingness to accept that life is inherently risky.

---

11   The following quotes are taken from the keynote address by Sir Harry Gibbs entitled "Living with risk in our society" that was delivered to a symposium organised by the NSW division of the Australia Academy of Technological Sciences and Engineering, Melbourne, 2002.

He is not alone in making these points. Social anthropologist Douglas observes: "We are ... almost ready to treat every death as chargeable to someone's account, every accident as caused by someone's criminal negligence, every sickness a threatened prosecution. Whose fault? is the first question" (Douglas, 1992:15, 16).

Gibbs is no doubt right when he talks of the rise of a culture of litigation, but his suggestion, and that of Douglas, that this represents a moral decline, misses an important point. When people suffer accidents, they may incur substantial financial loss. If a universal accident compensation scheme existed, people could apply to this scheme to make good the loss. New Zealand has such a scheme but Australia does not, so the only way that accident victims in Australia can make good their loss is if a court finds that some other party — preferably an insured party — is responsible. In short, however imperfectly, the law of negligence is filling a gap in the welfare system. The NSW Chief Justice neatly captured this situation in the title of a lecture that he gave in 2002, "Negligence: the last outpost of the welfare state" (Spigelman, 2002). I shall say more about this shortly. The point here is simply that the so-called litigious nature of modern society is better explained by current institutional arrangements than it is by the moral failings of modern individuals.[12]

Despite the analysis which Gibbs offers of the culture of litigation in terms of moral failure, he recognises that this culture could be combated by replacing negligence law with a no fault injury compensation scheme.[13] This would be fairer than the current system, he says, for at present, an injured person who can find someone to blame is compensated, while the person who can't, receives nothing.

There is, then, an irony here. The real reason for the culture of litigation is not the moral decline of the citizenry. Nor is it the ambulance-chasing lawyers. It is the judges themselves who have created this culture. Their purpose has been to compensate people for injury in circumstances where they would otherwise receive nothing and, to achieve this end, they have developed the fiction of the reasonable person who is considerably wiser than the ordinary traveller on the London Underground or, as one Australian judge put it, "the hypothetical person on the hypothetical Bondi tram" (Luntz and Hambly, 2002:244). By finding that duty holders have failed to comply with an idealised standard, judges have been able to ensure that the financial losses associated with injury are distributed and do not fall disproportionately on the injured party. The culture of litigation is an unintended consequence of this policy.

---

12   Another example of the institutional source of litigation can be found in the *Social Security Act 1991* (Cth). Sections 1166 and 1167 provide that a claimant may be compelled to take legal action against a partner or some other person as a pre-condition for receiving a benefit.

13   Such a scheme is recommended in the NSW Law Reform Commission report on a *Transport accidents scheme for NSW*, LRC 43/1, October 1984.

The welfare function of negligence law described above grew steadily from the 1960s to the 1990s but, from about 2000, the High Court called a halt to this trend and began to find in favour of defendants in a number of cases. The result, according to Spigelman, is that "the long-term trend has (now) been reversed" (Spigelman, 2002:433, 434).

Despite this reversal, Australia experienced an insurance crisis in 2002. Spigelman describes it graphically:

> "There were virtually daily reports about the social and economic effects of increased premiums: the abolition of charitable and social events, ranging from dances to fetes to surf carnivals, even Christmas carols; the closure of children's playgrounds, horse riding schools, adventure tourist sites, even hospitals; the early retirement of medical practitioners and their refusal to perform certain services, particularly obstetrics; the inability of other professionals to obtain cover for certain categories of risk led to similar withdrawal of services, for example, engineers advising on cooling tower maintenance could not get cover for legionnaire's disease, building consultants could not get cover for asbestos removal, agricultural consultants could not get cover for advice on salinity; many professionals were reported to have disposed of their assets so as to be able to operate without adequate, or even any, insurance" (Spigelman, 2003:293).

The crisis was, in part, a consequence of the pressure placed on insurance companies by the expanded welfare function of negligence law. However, as Spigelman points out, there were a number of other factors that brought matters to a head. In particular, there had been a number of disasters internationally, culminating in the events of 11 September 2001. The phenomenon of reinsurance, whereby insurance companies insure against losses that are so large they cannot afford to bare them alone, meant that these disasters took their toll on the insurance industry around the world. In Australia, the collapse of HIH (one of the country's biggest insurers) put further pressure on the industry. These factors, and others, drove premiums up far more rapidly than the gradual expansion of negligence law could ever have done. As Spigelman puts it, "2002 was the year in which quite a number of chickens came home to roost" (Spigelman, 2003:294).

The political response was rapid. The Commonwealth and state governments set up a high-level inquiry to review the law of negligence.[14] The inquiry recommended a number of changes, including changes to the way in which the concept of reasonable foreseeability was being interpreted by the courts, so as to bring the reasonable person of negligence law closer to the ordinary person in the street. These changes have been implemented legislatively around the

---

14 Commonwealth of Australia, Panel of Eminent Persons (Chairman: Ipp, D), *Review of the law of negligence report*, Canberra, 2002. Website at http://revofneg.treasury.gov.au/content/Report/PDF/LawNegFull.pdf.

country. Spigelman concludes, ". . . by a combination of major change in judicial attitudes, led by the High Court, and wide-ranging legislative change, the imperial march of the tort of negligence has been stopped and reversed" (Spigelman, 2003:311).

This is not the place to evaluate the desirability of these changes. However, it is worth drawing attention to the full significance of Spigelman's lecture title. Negligence law has indeed been an outpost of the welfare state. To the extent that the law abandons this outpost, and assuming that it is not replaced by a universal accident compensation system or some comparable institution, the welfare state will have given up significant territory, leaving citizens previously protected by the outpost to suffer the consequences of fate.

The reasonable person in OHS law

Whereas a major aim of negligence law (discussed above) is to ensure that injured parties are compensated, OHS law uses the concept of negligence to determine liability for punishment. Organisations must do what is reasonably practicable. Failure to do so is negligent, and punishable. Senior managers must exercise due diligence or reasonable care. Failure to do so is negligent, and punishable. In short, the civil law concept of negligence has become a basis for criminal liability.[15]

Consider, now, the consequences of this transfer from the civil to the criminal arena. In civil actions for damages, to say that one party is at fault for failure to exhibit exceptional foresight achieves the policy purpose of spreading the loss. It does not result in any social condemnation of or stigma against the party said to be at fault. The idea of fault here is somewhat artificial. But where such a finding of fault can result in punishment, particularly the punishment of an individual (as in the case of OHS law), matters are different. To say that a person is at fault for failure to exhibit exceptional foresight does indeed smack of injustice. To punish senior managers in these circumstances does seem unfair.

To put this in the context of the Gretley case, the reasonable manager and the reasonable surveyor, whom the judge in the Gretley prosecution had in mind, were people of unusual foresight. Ordinary managers and surveyors would probably have behaved in the same way that the Gretley managers and surveyors behaved when initially provided with the plans by the Department of Mineral Resources, that is, they would have accepted them as accurate. To punish the individual Gretley defendants for behaving in this way seems unfair. However, the situation was quite different in the days immediately prior to the accident when the indicators of danger began to emerge. At this point, the risk was not only reasonably foreseeable, it was foreseen. Had the prosecution focused on this, in the way that the earlier inquiry did, the divergence between

---

15 The concept of criminal negligence, which provides a basis for manslaughter prosecutions, is another matter. See Bronitt and McSherry (2005:183, 184).

the reasonable and the ordinary mine official would not have become the issue that it was. Individuals might then have been convicted without the sense of injustice that surrounded the judgments actually made.

To summarise, the comments made by the ACCI about holding individuals culpable for failing to live up to an idealised reasonable person standard have merit. There is indeed a need for some modification in the law in this respect, especially as it affects individuals. Just as the law of negligence has undergone recent legislative change, perhaps it is now time to modify OHS law to bring the reasonable person and the ordinary person more closely into line.[16] A more far-reaching legislative proposal will be discussed in a later chapter.

The reverse onus of proof

The campaign by the ACCI and other parties expressed great concern about the so-called reversal of the onus of proof in the NSW OHS Act. As noted earlier, the Institute of Public Affairs expressed outrage about this. Under the OHS Act, it said, managers "are presumed guilty until proved otherwise. Incredibly, they have to prove they are innocent". This, it was suggested, was contrary to the basic principles of criminal law. The ACCI policy paper included the following comment:

"Nor should there be a deemed guilt or reverse onus of proof in any civil or criminal proceedings for OHS breaches, nor any other basis on which employers, directors, management personnel or employees are treated less favourably than the defendants in prosecutions under any other equivalent law or legislation ... All parties charged with OHS offences should be accorded natural justice and, in criminal cases, the standard presumptions and protections of the general criminal law."[17]

There are two ways in which the NSW OHS Act reverses the onus of proof. The first is in relation to major duty holders, such as employers. These will normally be corporations. The second is in relation to individual senior officers of the corporation. I deal first with the corporate level.

A central provision of section 8 of the OHS Act is the following:

"An employer must ensure the health, safety and welfare at work of all employees of the employer."[18]

---

16 Some commentators argue that, in the case of corporate defendants, judges have already begun to rebalance the law in this way. See Bluff & Johnstone (2005:203).

17 Op cit, ACCI, p 46.

18 This is the wording in the *Occupational Health and Safety Act 2000* (NSW). The Gretley defendants were charged under the *Occupational Health and Safety Act 1983* (NSW). Section 15(1) of the OHS Act 1983 states that: "Every employer shall ensure the health, safety and welfare at work of all the employer's employees." The wording, in short, is virtually the same. For a full discussion of the differences between the two sections, see Thompson (2001:27ff).

It is for the prosecution to prove that there has been a failure to comply, and it must do so according to the normal criminal standard, beyond reasonable doubt. Once this is established, section 28 provides the following defence:

"It is a defence to any proceedings against a person for an offence against a provision of this Act or the regulations if the person proves that:

(a) it was not reasonably practicable for the person to comply with the provision, or

(b) the commission of the offence was due to causes over which the person had no control and against the happening of which it was impracticable for the person to make provision."[19]

It is subsection (a) that is relevant here. The defendant does not need to establish the case with the rigour that is imposed on the prosecution. It must prove its point only to the civil standard, that is, on the balance of probabilities (Thompson, 2001:91). In other words, it must show that its claim is more probable than not. This is not an onerous requirement. If it truly was not reasonably practicable for the defendant corporation to comply with the section, it should be relatively easy for it to establish the point. Nevertheless, in theory the situation involves a stark reversal of the onus of proof, with defendants required to establish their innocence.[20]

The reason for reversing the onus of proof in NSW appears to primarily be a practical one:

"[I]t is more efficient for the holder of the duty of care rather than the prosecution to have to establish what was reasonably practicable. A duty holder could be expected to know more about the costs and benefits of the various alternatives open to him or her at any time, than anyone else."[21]

The reverse onus of proof also applies in relation to certain personal defendants. Section 26 deems certain individuals to be guilty whenever the corporation is guilty, in the following terms:

"(1) If a corporation contravenes ... any provision of this Act ... each director of the corporation, and each person concerned in the management of the corporation, is taken to have contravened the same provision unless the director or person satisfies the court that:

(a) he or she was not in a position to influence the conduct of the corporation in relation to its contravention ..., or

---

19   This is the same as section 53 of the OHS Act 1983.

20   Queensland also reverses the onus of proof. All other Australian states have the same basic requirement that employers must maintain a safe workplace so far as reasonably practicable, except that it is for the prosecution to prove that it was reasonably practicable for the company to comply with the law. See Johnstone (2004:222).

21   Productivity Commission, *Work, health and safety*, Report No 47, September 1995, p 55 (quoted in Thompson (2001:93)).

(b) he or she, being in such a position, used all due diligence to prevent the
       contravention. . ."

The fact that individuals are deemed guilty in this way means that there is no
need for the prosecution to prove anything, except that the person was
concerned in the management of the corporation. This must be proved beyond
reasonable doubt. In the Gretley case, managers and surveyors were found to be
persons concerned in the management of the corporation, but the prosecution
failed to convince the judge that shift managers were such persons.

Section 26 also provides a defence. The relevant element is that the defendant
used *all due diligence* to prevent the offence by the corporation, that is, exercised
an appropriate level of care. This is for the defendant to prove but, again, he or
she needs only do this on the balance of probabilities to avoid conviction. In
principle, this should be relatively straightforward, if the defendant did indeed
use all due diligence. Certain other Australian jurisdictions also reverse the onus
of proof in this way, so NSW is not unique in this regard.[22] Again, the
justification for the reversal would seem to be that it is easier for the defendant
than the prosecution to present the necessary information.

It is clear from the above discussion that the phrase "reverse onus of proof" is
somewhat misleading. It suggests a transfer to the defendant of the burden of
proof that is normally carried by the prosecution. The situation is not
symmetrical, however. The defendant does not have to prove innocence as
conclusively as the prosecution must normally prove guilt.

In any case, it can be argued that the matter is of little practical significance — at
least it appeared not to be in the Gretley case. This is because the Gretley trial did
not work in the way formally envisaged in the Act. Consider, first, the corporate
defendants. The judge did state formally that "the defendants have failed to
discharge the onus placed upon them" of proving that it was not reasonably
practicable to ensure safety.[23] But the reality was that the prosecution had not
simply left it to the defence to establish this point. It had shouldered the onus
itself and taken active measures to prove that it *was* reasonably practicable for the
companies to have avoided the breach. It called two expert witnesses who
described what reasonable managers and surveyors would have done.
Reasonable surveyors would not have relied on the plans provided by the
department and reasonable managers would have probed their surveyors'
actions more carefully. The judge accepted this evidence.[24] Furthermore, she
found that the evidence provided by the prosecution demonstrated that the
defendants had not behaved in accordance with this model of reasonable
behaviour.[25] In short, the prosecution itself established to the satisfaction of the

22   Queensland and Tasmania; see Johnstone (2004:434, 435).

23   J 824.

24   J 801-804.

25   J 811.

judge that the companies had not done what was reasonably practicable. It did not simply assume reasonable practicability and leave it to the defence to prove otherwise; it actively set out to prove that the companies had not done what was reasonably practicable and the judge found the prosecution evidence to be compelling. It can be concluded that, even if the Act had formally placed the onus on the prosecution to establish reasonable practicability, the outcome of the trial would not have been very different in relation to the corporate defendants.

The situation was much the same with respect to the personal defendants. The discussion above treats the behaviour of the surveyors and managers as the behaviour of the companies. This is in accordance with the established principle that "a corporate employer can only conduct its activities through human agents such as managers, supervisors, employees and contractors" (Johnstone, 2004:206, 229). The company was convicted, in part, on the basis of the behaviour of these individuals. In the Gretley case, this same behaviour resulted in charges against the individuals as well. The onus of proof was formally on them to establish that they had exercised all due diligence. The judge found that they had failed to produce any such evidence. On the contrary, the evidence produced by the prosecution established that they had *not* exercised all due diligence.[26] As in the case of the companies, even if the onus had been on the prosecution to prove that these individuals had failed to act with due diligence, the outcome would probably have been much the same. In the Gretley case, the issue of where the onus of proof lay turns out to have been a red herring.

The reverse onus in criminal law

The critics assert that reversing the onus of proof, as is done in OHS law, is contrary to the basic principles of criminal law. The fact is, however, that whatever the principles of criminal law may be, in practice, there are many instances where the onus of proof is reversed in order to achieve policy/regulatory objectives. It is worth outlining some of these instances in order to put the debate about the reverse onus of proof in a broader context.

The NSW *Summary Offences Act 1988* makes it an offence to live on the earnings of prostitution. How can the prosecution prove that a person who lives with a prostitute is living on the earnings of prostitution? The issue is resolved by *assuming* that this is the case, unless the defendant can show that he or she has lawful means of support (see the legislation in the following box). In this way, the onus of proof is reversed and the defendant is required to establish his or her innocence. Notice, moreover, that the offence is a serious one in that the maximum penalty is 12 months in gaol. Even so, the law is willing to reverse the onus of proof in order to ensure that the provision can be effectively enforced.

---

26   J 963, 964.

Summary Offences Act 1988, section 15

15(1) A person shall not knowingly live ... on the earnings of prostitution of another person.

Max penalty: 10 penalty units or imprisonment for 12 months.

15(2) For the purposes of subsection (1), a person who ...

(a) lives with or is habitually in the company of, a reputed prostitute, and

(b) has no visible lawful means of support,

shall be taken knowingly to live ... on the earnings of prostitution ... unless he or she satisfies the court ... that he or she has sufficient lawful means of support.

A second example from the same Act concerns the possession of knives in public places, particularly in schools (see the legislation in the following box). The provision makes it an offence to carry a knife unless the person concerned can provide a reasonable excuse. Here again, the onus is on the accused to prove his or her innocence. And here again, the offence is not trivial: it can lead to 12 months' imprisonment.

Summary Offences Act 1988, section 11C

"11C(1) A person must not, without reasonable excuse (proof of which lies on the person), have in his or her custody a knife in a public place or a school.

Maximum penalty: ... 5 penalty units or, ... in the case of a person dealt with previously for a knife-related offence — 10 penalty units or imprisonment for 12 months, or both.

Consider, next, drug law. The law normally treats possession of a drug for the purpose of sale as a more serious offence than possession for one's own use. But how can the prosecution establish that the accused intended to sell the drugs found in his or her possession? In NSW, the problem is solved by simply assuming that any quantity of drugs above a certain threshold is intended for sale (see the legislation in the following box). However, the accused is given the opportunity to prove that this was not the case, for example, by proving that the drug was for personal use.

Drug Misuse and Trafficking Act 1985, section 29

A person who has in his or her possession an amount of a prohibited drug which is not less than the traffickable quantity of the prohibited drug shall ... be deemed to have the prohibited drug in his or her possession for supply, unless:

(a) the person proves that he or she had the prohibited drug in his or her possession otherwise than for supply, or. . .

The onus of proof is also reversed for financial offences committed by company directors. Where a company is insolvent, that is, unable to pay its debts, directors have a duty to prevent the company from incurring further debts. If a company does incur a debt while insolvent, then, under Commonwealth law, its directors are guilty of an offence unless they can prove that they had reasonable grounds to think the company was solvent or, if they suspected that it was insolvent, that they took reasonable steps to prevent it trading (see the legislation in the following box).

Corporations Act 2001, section 588H

588H(2) It is a defence if it is proved that, at the time when the debt was incurred, the person had reasonable grounds to expect, and did expect, that the company was solvent at that time . . .

588H(5) It is a defence if it is proved that the person took all reasonable steps to prevent the company from incurring the debt.

These examples could be multiplied.[27] They show that OHS law is not in any way unique in placing the onus of proof on defendants to prove certain things in order to avoid conviction. The critics claim that the reversal of the onus of proof under the OHS Act is unjust. If that is so, many other statutes are unjust in the same way. The fact is that effective law enforcement relies on placing the onus of proof on defendants in various circumstances. Governments have decided at times that holding senior officers personally liable for safety offences committed by their companies, unless they can show that they exercised due diligence, is an effective way of encouraging corporate compliance. The critics of this policy are apparently unaware that the reverse onus is to be found in many parts of criminal law.

Absolute liability

The ACCI policy statement was very critical of the absolute liability imposed on employers by the NSW OHS Act. An absolute duty, it says, is "hostile to the common law intent and to common sense".[28]

What does absolute liability mean? An offence is one of absolute liability if it is defined independently of any question of fault or blameworthiness.[29] For example, motorists who exceed the speed limit are guilty of an offence, regardless of whether they realise that they are doing so (Bronitt and McSherry,

---

27   For example, the *Protection of the Environment Operations Act 1997* (NSW), section 118; the *Environment Protection Act 1997* (ACT), section 153; and the *Trade Practices Act 1974* (Cth), section 85.

28   Op cit, ACCI, pp 41, 45.

29   Absolute liability offences take no account of whether the defendant has made an honest and reasonable mistake of fact. Where the defence of honest and reasonable mistake of fact is available, the offence is described as one of strict liability. See Bronitt and McSherry (2005:187ff).

2005:189). For the prosecution to have to prove that motorists knew what the limit was and knew that they were exceeding it would make the law unenforceable.

Absolute liability is not limited to what might be described as regulatory offences, such as exceeding speed limits. There are absolute liability offences in traditional criminal law — offences that can result in long terms of imprisonment. Suppose a person intentionally inflicts serious harm on another, but without intending to kill the victim or foreseeing that the victim might die. Suppose the victim subsequently dies. The attacker is then absolutely liable for the death and is guilty of murder even though, in a significant sense, the death was accidental.[30] Like it or not, absolute liability is a feature of criminal law.[31]

To return to the ACCI policy statement, it quotes the following judicial comment on the absolute nature of the NSW OHS Act:

> "The duties imposed by the Act are not merely duties to act as a reasonable or prudent person would in the same circumstances . . . Under s15(1) the obligation of the employer is 'to ensure' the health, safety and welfare of employees at work. There is no warrant for limiting the detriments to safety contemplated by that provision, to those which are reasonably foreseeable . . . the terms of s15(1) specify that the obligation under that section is a strict or absolute liability to ensure that employees are not exposed to risks to health or safety."[32]

From an employer's point of view, this is a horrifying statement. It creates the understandable impression that the law will hold employers liable and punishable in the event that an employee is injured or killed, regardless of whether the employer is at fault.

This is an incorrect impression, and it is worth commenting on how this misconception comes about. There is a fundamental distinction in criminal law between, on the one hand, the elements of an offence that the prosecution must prove and, on the other, matters which it is for the defendant to establish. These different components are often widely separated in the legislation, with the elements that the prosecution must prove towards the beginning of the Act and the defences available to the defendant spelt out in sections towards the end. Nevertheless, these different components need to be read together to understand the nature of the offence. When applying this to the NSW OHS Act, the offence, as a whole, is failure to maintain a safe workplace so far as reasonably practicable. The prosecution's task is to prove that the defendant breached its duty to maintain a safe workplace. It does not have to consider whether the defendant is at fault. In this particular sense, the duty is absolute. It is for the

---

30 Bronitt and McSherry (2005:469). Fisse (1990:56) describes this as a case of strict liability.

31 Various commentators have recommended the abolition of this form of murder. See Bronitt and McSherry (2005:470).

32 Op cit, ACCI, p 43.

defendant to establish, if possible, that it was not reasonably practicable to avoid the breach. In other words, it is for the defendant to establish that it[33] was not at fault, always remembering that this should be done only on the balance of probabilities (Johnstone, 2004:221) — not an especially onerous burden. The offence *as a whole* is thus not one of absolute liability; a defendant will only be found guilty if the breach was indeed its fault. To repeat, the duty to maintain a safe workplace is absolute only in the sense that the prosecution does not need to establish fault in order to find that the duty has been breached, but a conviction is only possible after considerations of fault have been canvassed.[34]

It can be seen that the statement quoted above by the ACCI is only half the story. It refers only to section 15(1), which spells out the duty, and says nothing about the defence provided at the end of the Act in section 53.[35] This is not to criticise the ACCI. Its omission is entirely understandable. The architecture of the OHS Act is thoroughly confusing to those who have not studied it closely, and judges contribute to the confusion by making statements about the duties imposed on defendants without making reference to the defences available to them.[36] But the upshot of this discussion is that the concerns expressed by the ACCI and others about absolute liability in the NSW OHS Act are ill-founded.

Evaluating the arguments: a summary

The campaign triggered by the Gretley convictions centred on three specific concerns about the way in which the NSW OHS Act has been formulated and interpreted: the legal concept of reasonableness, the reverse onus of proof, and the absolute nature of the duty of care. I have argued that the critics are right in pointing to the fact that the reasonable person envisaged in the law is not the ordinary person and that to punish people for failing to live up to an ideal can create a sense of injustice. Holding senior managers accountable for the health and safety performance of their organisation is entirely appropriate, but it is not always appropriate to assume that they are at fault when the organisation fails to manage risk effectively. Later chapters will explore ways of holding senior managers accountable without assuming personal fault.

---

33   For ease of expression here, I assume a corporate defendant.

34   According to Parry (1995:695), where a due diligence defence is available, liability is "strict" but not "absolute". But see Duncan and Traves (1995, ch 8).

35   These section numbers are from the OHS Act 1983. The corresponding section numbers in the OHS Act 2000 are 8(1) and 28.

36   Brooks (1993:496) makes the same point: "*Carrington Slipways* says the duty is absolute, and that sec 53, with its common law implications, is only a defence. But in terms of the average employer contemplating his or her advisable course of action, such statements are arid legalisms. What counts in not what is the duty and what the defence, but 'What actions can be taken which will not attract the threat of liability?'"

The other two concerns — the reverse onus of proof and the absolute nature of the duty of care — have been shown to be of little substance. The offence as a whole does not impose absolute liability and the reverse onus of proof does not constitute a problem, in and of itself. The defendants in the Gretley prosecution were unable to demonstrate that they had acted reasonably, but this was not because they bore the onus of proof. It was because they had not in fact acted reasonably, according to the legal definition of reasonableness. Had the legal notion of reasonableness been more in accordance with common usage, the defendants in the Gretley case would not have had much difficulty establishing that they behaved reasonably in relation to the plans provided by the Department of Mineral Resources. However, even using a commonsense meaning of reasonableness, it is far from clear that the response to the warning signs was reasonable. Had the prosecution highlighted these issues, the defendants would have been hard-pressed to establish that they had acted reasonably, regardless of which meaning of the term was used.

The outcome of the campaign

The campaign mounted by business interests against the NSW OHS Act appeared to meet with considerable success. In mid-2006, the government produced a draft Bill to amend the Act.[37] However, there was strong opposition to the draft and the government chose to seek further legal advice before proceeding. At the time of writing, the outcome remains in doubt. Among other things, the Bill addressed the three matters raised here: reasonableness, the reverse onus of proof, and the absolute duty of care. It is worth considering briefly how the Bill intended to deal with these issues.

Absolute duty of care and reverse onus of proof

The draft Bill proposed that employers would need to ensure health and safety, but only so far as was reasonably practicable. This formulation abolished the seemingly absolute nature of the duty imposed on employers, since that duty was now clearly qualified by reasonable practicability. In so doing, the Bill shifted the onus of proof back to the prosecutor, who would now have the responsibility of proving that it was not reasonably practical for the employer to ensure the health and safety of people in the workplace. In these matters, then, the Bill reflected employers' concerns.

Some commentators claimed that shifting the onus of proof back to the prosecution in this way represented a "big change to a commonsense approach."[38] However, it was shown earlier that, regardless of the formal

---

37 NSW Occupational Health and Safety Amendment Bill 2006, 3 May 2006.

38 According to one legal firm, "this change in onus *may* result in a lesser number of prosecutions, as prosecutors will need to be satisfied before instigating a prosecution that they can prove each element of the offence, including the steps which were reasonably practicable for the defendant to have taken" (emphasis added) (statement by Freehills, mid-2006, reproduced in various trade publications).

situation, the prosecution in the Gretley case shouldered the responsibility for demonstrating that the defendants had not done what was reasonably practicable. It follows that the change foreshadowed in the draft Bill would not have saved the Gretley defendants from conviction.

Reasonableness

The Bill also attempted to deal with the issue of reasonableness. It provided guidance on matters to be taken into consideration when deciding what is reasonably practicable,[39] but it failed to address the fundamental problem identified above, namely, the way in which the reasonable person has diverged from the ordinary person. There was nothing in the guidance that would have saved the corporate defendants in the Gretley case from conviction.

In relation to individual defendants, the amendments removed the requirement for "due diligence" and replaced it with a requirement for "reasonable care". On the face of it, this does not seem like a significant change.[40] Again, there was nothing in the Bill that required courts to interpret reasonableness in light of ordinary behaviour. It could be that the change in judicial thinking referred to earlier will bring about the intended effect, but the Bill itself seemed to miss the mark in this respect.

It appears, then, that the changes proposed in the draft Bill were not as significant as they appeared at first sight. Whether or not the new provisions are eventually enacted, a Gretley style prosecution will remain a possibility — unless judges themselves decide that the concept of reasonableness that they are working with is not itself reasonable.

---

39 This is essentially the guidance given in the Victorian *Occupational Health and Safety Act 2004*.

40 The amendments also contain guidance about things to be taken into account when deciding what is reasonable care. The list includes "what the officer knew about the matter". The Freehills commentary referred to earlier states that this is "aimed at ensuring that an officer will only be held liable for matters which the officer knew about …". Ignorance, in short, would be an excuse. Such an interpretation makes no sense, however, for the question is not what the defendant knew, but what the reasonable person would have known.

Chapter 6
Industrial manslaughter

The public pressure that led to prosecution in the Gretley case was part of a wider demand for industrial manslaughter law. The CFMEU (the mining union) in fact argued that the Gretley case demonstrated the need for such legislation in order to more effectively prosecute employers in industrial fatality cases in the future.[1] So-called industrial manslaughter legislation has been enacted in various Australian jurisdictions, and the conviction of the Gretley defendants has led some company managers to believe that they are now at risk of conviction under these new laws. This is far from the case. It will be useful to describe this legislative development here, both to distinguish it clearly from mainstream OHS legislation under which the Gretley prosecutions occurred and to show how limited the effect of these provisions is likely to be. I shall argue that these new enactments are largely symbolic in nature.

The pressure for industrial manslaughter law

There are two main sources of pressure for industrial manslaughter law. The first has come from victim and employee groups arguing for legislation that will enable company officers to be sent to prison when people are killed as a result of their negligence. Following the 1998 Longford gas plant accident in Victoria and the subsequent prosecution of Esso Australia, a union spokesman claimed that "Esso's senior officers should have been faced with charges of manslaughter" (Hopkins, 2002:49). In NSW, unions have placed enormous pressure on the Labor Government to enact laws which would allow the jailing of employers who fail to protect the lives of workers. In 2003, for example, around 10,000 workers marched through Sydney streets demanding tougher laws following the death of a 16-year-old who fell from a building site roof on his third day at work as a roof plumber's labourer.[2]

There are also international parallels. In Canada, the Westray mine disaster in 1992 (which killed 26 men) generated persistent union pressure for industrial manslaughter legislation (Glasbeek, 2005:40,42). In the UK, a series of mass transport disasters — including a capsized ferry and several train crashes — saw the formation of various groups representing the families of victims, and these groups have pressed relentlessly for manslaughter legislation so that the individuals whom they believe to be responsible can be brought to justice (Gobert, 2005:21).

---

1   ABC, 5/11/04.

2   AAP, 30/10/03.

The second source of pressure for industrial manslaughter law has come from academics and other legal commentators arguing for the need to hold *corporations* accountable for the deaths that they cause (Fisse, 1990; Wells, 2001). Attempts to charge corporations with manslaughter in the past have been thwarted because criminal law has traditionally focused on individual criminality, making it difficult to talk about the criminality of corporations. Legal commentators have long suggested that manslaughter laws need to be changed to reflect the reality of corporate negligence. They argue that decision-making in large corporations is diffuse, and that courts need to be able to aggregate the negligence of individuals at various levels of the company in order to be able to convict a corporation of manslaughter. These same commentators argue that the culture of a corporation may be a culture of carelessness or a culture of risk denial, and that such cultures need to be taken into account when determining corporate liability. Of course, corporations convicted of manslaughter could not be imprisoned, but they might be subjected to far heavier fines than are possible under ordinary OHS law.

As this discussion shows, advocates of industrial manslaughter law have two quite different types of target in mind. The relatives and workmates of the victims want the law to target individuals and, where appropriate, send them to jail, while legal commentators have tended to focus on the culpability of the corporation itself and to argue specifically for *corporate* manslaughter law.

Degrees of culpability

A crucial question, perhaps *the* crucial question, concerns the degree of culpability that would be necessary to warrant a conviction. In the case of individuals, this is a critical issue, since they face the possibility of imprisonment should they be found guilty. In the case of corporations, it is perhaps not as critical, since they face only the possibility of heavier fines. Nevertheless, the leaders of large corporations do not relish the prospect of their companies being found guilty of corporate manslaughter because the stigma affects them as well as their company.

In order to understand the legislative approach to this issue, we must begin by identifying three levels of culpability, in ascending order.

Civil negligence

For the most part, the degree of culpability required for the conviction of an individual or a corporation under OHS law is failure to do what the reasonable person (or corporation) would do. This is the so-called civil standard of negligence. As has been shown in previous chapters, there is plenty of room for argument as to how the reasonable person would behave, but that is not the issue here. The point is simply that a defendant is rendered liable by any failure to behave as the reasonable person would behave.

Gross negligence

Gross negligence is one step up on the culpability scale. It is also described as criminal negligence (to distinguish it from civil negligence) but, since civil negligence can lead to a criminal conviction (as it does in the case of OHS offences), describing this higher level of negligence as criminal can be more confusing than helpful.[3] The real significance of gross negligence is that it is the basis for a manslaughter conviction under traditional criminal law and will attract heavier penalties than ordinary or civil negligence. In particular, it is the kind of negligence that might warrant imprisonment.

The question, then, is how serious does negligence have to be before it can be described as gross? In terms of the reasonable person standard, how far short of the standard must the behaviour fall before it can properly be described as gross? This is a crucial question, since people's liberty is at stake.

Astonishingly, the law provides no coherent answer. In Australia, all discussions of this point go back to a 1977 ruling by the Victorian Supreme Court. It stated that an act that caused death was manslaughter if it was done:

> "... in circumstances which involved such a great falling short of the standard of care which a reasonable man would have exercised and which involved such a high degree of risk that death or grievous bodily harm would follow that the doing of the act merited criminal punishment."[4]

There is a fundamental circularity here. What we are looking for is a statement of the characteristics of negligence that would justify describing it as gross and therefore meriting criminal punishment. Instead, we are told that if the negligence merits criminal punishment, it can be described as gross. Essentially, what is being said is that, if the behaviour is judged to merit criminal punishment, then it merits criminal punishment. There is no guidance

---

3   Occupational health and safety law is generally regarded as criminal law in that it imposes significant penalties and the prosecution must prove its case beyond reasonable doubt. Accordingly, it may appear somewhat anomalous that the level of culpability required for conviction is the civil level rather than the so-called criminal level. However, since individual offenders cannot generally be sent to jail, particularly for a first offence, legislators have judged that this inconsistency does not in itself amount to an injustice. It should be remarked that inconsistency is not in itself a ground for questioning the legitimacy of some law or judgment. One criminal law text explains the situation as follows: "The apparent instability of fault in the criminal law may simply be a reflection of the diversity of functions performed by the modern criminal law ... Our conclusion may simply be that the meaning of fault in the criminal law is historically contingent, and that any quest for the 'grand universal theory' to account for all forms of culpability may prove to be illusive, if not illusory." See Bronitt and McSherry (2005:186).

4   *Nydam v R* [1977] VR 430, at p 445.

whatsoever here for decision-makers, be they judge or jury. It is up to decision-makers to decide for themselves whether the behaviour merits criminal punishment — drawing, perhaps, on broader community standards.[5]

It is clear, then, why employers might fear industrial manslaughter charges that are based on the test of gross negligence, since it is really no test at all. The only question would be whether the judge or jury thought that the behaviour was sufficiently serious to warrant criminal punishment, in particular, the possibility of imprisonment. Given that the community is increasingly intolerant of workplace fatalities, there could well be an increasing range of circumstances in which behaviour might be judged as grossly negligent.

Recklessness

Recklessness is a third, and yet higher, level of culpability. According to most authorities, a person's conduct is reckless if he or she foresees the likelihood of a harmful consequence occurring, but nevertheless engages in the conduct (Bronitt and McSherry, 2005:179). The crucial point here is that the person actually foresees the consequences. No longer is it a question of what the reasonable person might foresee; it is what was actually in the mind of the defendant. The job of the prosecution is not to establish that the reasonable person would have foreseen a certain outcome, but that the defendant actually foresaw it. This provides far greater protection to defendants against any charge of reckless behaviour than is available against a charge of negligent behaviour, be it gross or otherwise. If the defendant did not foresee the risk, then that is just about the end of the matter.[6]

Moreover, what must be foreseen is not a remote possibility, but a likelihood. Thus, although on one interpretation the Gretley defendants in the days immediately before the accident did foresee the possibility of what might happen, there was no evidence that they saw this as a likely outcome. There could be no suggestion, therefore, that their behaviour was reckless.

Recklessness sits close to the top of the culpability scale. From a philosophical point of view, it is even worse to commit an offence intentionally. However, in the eyes of the courts, to cause death as a result of reckless behaviour is just as blameworthy as it is to cause death intentionally (Bronitt and McSherry, 2005:180). Recklessness can therefore be the basis of a charge of murder.

---

5 Juries have frequently failed to uphold manslaughter charges against drunk drivers who have killed someone because, in the absence of clear guidance as to the necessary level of culpability, they have tended to view such behaviour as insufficiently serious to count as gross negligence. For this reason many jurisdictions now charge such offenders with "culpable driving causing death", a charge which jurors seem more willing to uphold. See Bronitt and McSherry (2005:485).

6 But see "Caldwell recklessness", Bronitt and McSherry (2005:182, 183).

The new industrial manslaughter provisions

Let us consider now some of the so-called industrial manslaughter provisions that have been enacted in Australia. In so doing, I shall be considering two questions. First, what level of culpability is envisaged as a prerequisite for conviction? Second, do the provisions target individuals, as victim groups want, or corporations, as legal commentators want, or both? I shall restrict attention here to three jurisdictions: NSW, Victoria and the ACT.[7]

NSW

There has been pressure for industrial manslaughter legislation in NSW for some years. In 2002, the Australian Democrats introduced a Crimes (Corporate Manslaughter) Bill. The ALP also declared that its policy was to enact an offence of "industrial manslaughter". However, the Labor Government was generally reluctant to introduce the new offence, believing that current manslaughter law was adequate. Indeed, a government spokesman said at one stage that the government would not enact industrial manslaughter law.[8]

However, the pressure continued, culminating in 2005 with the insertion into the OHS Act of a new provision concerning workplace death. The ministerial statement that accompanied the Bill claimed that the new provision was aimed at a small minority — the rogues who recklessly endangered the lives of others. The vast majority of employers had nothing to fear.[9] The minister stressed that the business community supported the government's position. He quoted a survey undertaken by the state Chamber of Commerce that showed that "89.7 per cent of businesses surveyed believed employers who 'deliberately and recklessly' put their employees' lives at risk should be gaoled".

Crucially, the level of culpability specified in the new provision was recklessness. In other words, conviction would depend on whether the prosecution could show that a person was aware that the conduct in question was likely to cause death.[10] Recalling the discussion above, it would seem that this was not an industrial manslaughter provision, but an industrial murder provision. Only if the level of culpability justified a murder charge would defendants be liable under this provision. But, despite the seriousness of the conduct that was targeted by this provision, the maximum possible penalty was five years in prison — hardly an excessive penalty for murder.

---

7   Other Australian jurisdictions either have less onerous industrial manslaughter provisions or no provisions at all.

8   AAP, 7/10/03, 29/10/03, 30/10/03.

9   Ministerial statement accompanying the introduction of the Occupational Health and Safety Amendment (Workplace Deaths) Bill 2005.

10   Foster (2006:79-81) suggests that the courts might interpret the provision as requiring only that the defendant foresaw the possibility, not the probability, of the result.

Union groups expressed qualified support for the new provision.[11] They were critical of the relatively low maximum penalties, but they seemed unaware of how difficult it would be to obtain convictions. Although some commentators saw the new provision as a victory for employee groups, one would have to say that it was a symbolic victory only. Prosecution under this provision will be possible in only a tiny minority of cases where workers are killed. In many cases, it will be easier to prosecute for manslaughter using general criminal law rather than the new statutory provision.

With regard to the question of who or what the provision targeted, it certainly targeted individuals, including directors, managers and fellow workers — whomever could be shown to have behaved in a reckless fashion — but it also targeted corporations. The minister noted that "a corporation that engages in reckless conduct that causes the death of a person at a workplace can also be charged under the new provision".

However, large corporations in NSW have little to fear. I noted earlier that the bias in criminal law towards individual rather than corporate liability made it difficult to prosecute corporations for manslaughter. The NSW provision does nothing to address this issue. It is likely to prove even more difficult to attribute recklessness to corporations than it has been to attribute gross negligence. The failure of the workplace death legislation to deal with this issue reinforces the conclusion that the enactment was largely symbolic in nature.

Victoria

The Victorian *Occupational Health and Safety Act 2004* contains a provision that has been described as an industrial manslaughter provision.[12] Section 32 states, in part, that:

> "A person who, without lawful excuse, recklessly engages in conduct that places or may place another person who is at a workplace in danger of serious injury is guilty of an indictable offence and liable to . . . a term of imprisonment not exceeding 5 years."

As in NSW, the culpability test is recklessness, which severely limits the circumstances in which the provision may come into play. Interestingly, the provision is not limited to cases where there is a fatality, although that is where it is most likely to be used, if ever. Finally, the provision is theoretically applicable to corporations but, as in NSW, it fails to provide any means for establishing that corporations were reckless. In short, the provision appears to be a largely symbolic gesture in the direction of union concerns.

Section 32 is a far cry from the Bill that was put to the Victorian Parliament in 2001 but defeated in the Upper House in 2002 (Hopkins, 2002). That Bill focused in the first instance on corporate rather than individual defendants, and it

---

11   AAP, 6/5/05.

12   See Australian Mines and Metals Association, National Circular No 47/2005, 29 June 2005.

required the prosecution to show gross negligence, not recklessness. Moreover, it specified how the courts were to go about assessing whether a corporation was grossly negligent. It provided that "the conduct of any number of employees, agents or senior officers of the body corporate may be aggregated" in order to establish the negligence of the corporation. The Bill also went into considerable detail about the kinds of corporate failures that would count as evidence of corporate negligence. This was not a symbolic gesture but a serious and considered attempt to hold corporations responsible for manslaughter.

Although the 2001 Bill was aimed at corporations, it encompassed individuals as well. It provided that if, and only if, a corporation was found guilty of manslaughter, senior officers might also be convicted, but only if they were reckless, that is, only if their behaviour amounted to murder. In an earlier draft of the Bill, the offence had been one of gross negligence, but business lobbying was successful in limiting liability to situations of recklessness. This Bill was therefore no more of a threat to individual defendants than the provisions currently in place in NSW and Victoria. At the time, however, it was unacceptable to the Victorian Upper House.

The ACT

The ACT is the only jurisdiction in which thoroughgoing industrial manslaughter legislation has been enacted. The level of culpability envisaged is either recklessness *or* gross negligence.[13] The fact that it refers to gross negligence makes this truly manslaughter legislation. The maximum possible penalty for individuals is 20 years imprisonment (presumably, where the level of culpability is gross negligence rather than recklessness, the effective maximum will be considerably lower).

This legislation targets corporations as well as individuals. A corporation can be found guilty of reckless behaviour if it is proved:

"... that a corporate culture existed within the corporation that directed, encouraged, tolerated or led to noncompliance ... or ...

that the corporation failed to create and maintain a corporate culture requiring compliance ..."[14]

Moreover, in order to establish the gross negligence of a corporation, the court can aggregate the conduct of a number of its employees, agents or officers.[15]

---

13   The *Crimes (Industrial Manslaughter) Amendment Act 2003* (ACT) speaks only of negligence, but the Act must be interpreted in the light of the *Criminal Code 2002* (ACT), section 21, where negligence is defined as gross, as in *Nydam* (see fn 4). The offences defined in the ACT Act are modelled on the fatal offences provisions in ch 5 of the *Model Criminal Code* of the Standing Committee of Attorneys-General.

14   *Criminal Code 2002* (ACT), section 51(2).

15   *Criminal Code 2002* (ACT), section 52(2).

These provisions overcome the previously mentioned individualistic bias in criminal law and make it realistically possible to charge corporations with manslaughter in appropriate circumstances.

Discussion

The unions in Australia have pressed hard for industrial manslaughter legislation in the hope that, when workers are killed, employers will be liable to imprisonment in appropriate circumstances. They have not, however, paid close attention to the question of what those appropriate circumstances might be. In particular, they have not been clear about the level of culpability that might warrant a term of imprisonment. Yet this particular question has been paramount when it comes to drafting industrial manslaughter provisions. In NSW and Victoria, legislators have opted to impose liability only when employers stand accused of the most extreme type of culpability imaginable in a workplace context — recklessness. The vast majority of employers have nothing to fear from such laws; as its proponents frequently say, it is aimed at rogues, not honest men and women. In addition, such legislation is not likely to result in prosecutions of significant companies. In the absence of any legislative guidance as to how corporate recklessness might be demonstrated, it will normally be impossible to establish that a company was reckless. Perhaps only in the case of the smallest, "one-man" companies will it be possible to attribute the recklessness of an individual to the company and thus, in certain circumstances, to obtain a corporate conviction.

The ACT legislation is the only serious industrial manslaughter legislation in Australia. It is the only legislation that specifies how the culpability of corporations might be demonstrated, and it is the only legislation that envisages holding individuals accountable for gross negligence — the traditional level of culpability for manslaughter.

The movement for industrial manslaughter legislation is primarily fuelled by the union demand for some form of personal accountability when workers are killed. It has been bitterly opposed by employer groups who argue that, generally speaking, the culpability of employers when workers are killed is not such as to warrant manslaughter charges. Legislators have tried to respond to both sets of concerns and, in NSW and Victoria, they have enacted legislation that will be inapplicable in almost all cases of workplace death and thus of no assistance to relatives seeking accountability and closure. As if in recognition of this, a NSW Government spokesmen said at one point that the government would develop a system of grief counselling for relatives of people killed at work.[16] This will not, of course, provide them with what they want.

---

16   AAP, 29/1/03.

From the point of view of grieving relatives, industrial manslaughter legislation to date has been a largely futile exercise. It is harder to convict individuals under the NSW and Victorian provisions than it is to gain manslaughter convictions under general criminal law, and even in the case of the ACT legislation, it is probably no easier. This is not to say that general criminal law can be expected to provide much comfort for the bereaved. There have been very few attempts under general criminal law to bring manslaughter prosecutions for workplace deaths in Australia, and these few cases have been largely unsuccessful.[17] The next chapter of this book will suggest a radically new way in which the needs of grieving relatives can, to some extent, be met.

Let me conclude on a practical note. The Gretley trial in NSW and the Esso trial in Victoria were both showcase trials, and both stimulated union demands for industrial manslaughter legislation. Such legislation has since been enacted in both states but, even if it had been in place at the time of these tragedies, there is virtually no chance that any of the companies concerned, or their managers, would have faced charges under these provisions.[18]

---

17   Johnstone (2004:475). There have been more successful cases in the UK; see Foster (2006:87, 88).

18   For a discussion of the Esso case, see Hopkins (2002). The Esso case is also discussed by Foster (2006:91).

Chapter 7

Holding corporate leaders responsible

Occupational health and safety law consists of a combination of corporate and individual responsibility. There are good reasons for this. Offences are often unequivocally corporate in nature, in the sense that it is easy to see how the company as a whole has failed, while at the same time it is often difficult, if not impossible, to sheet home the failure to a single individual (Fisse and Braithwaite, 1993). In these circumstances, it makes sense to prosecute the corporate entity. On the other hand, corporations act through human agents and there is a strong argument that identifying these agents and holding them responsible is one of the most effective ways to ensure corporate compliance (Cressey, 1989). Hence, OHS law often contemplates holding responsible those people who take part in the management of the corporation. Obviously, the higher in the corporate hierarchy these managers are, the better. Chief executive officers and company directors are the people with the greatest capacity to influence corporate behaviour, so holding these people responsible for corporate failures provides the greatest potential leverage.

At this point, the issue of fault becomes critical. Current legislation holds individuals liable for corporate fault only if they were themselves individually at fault in some way, for example, by failing to exercise due diligence. The state of NSW has been more active than others in prosecuting directors who were personally at fault, but a recent study by Foster has shown that nearly all of these cases involved very small companies — small family concerns or even one-person companies — and that, in nearly all cases, the director was personally involved in the incident (Foster, 2005). In an important sense, these are not cases in which a director has contributed to an offence by a corporation but, rather, cases in which the offence is wholly and solely attributable to the director.

The Gretley prosecution was an attempt to break out of this mould in that it targeted individual managers in a much larger corporate structure. However, although the mine managers were the most senior people on site, they were separated by layers of management from the top corporate decision-makers. In short, the Gretley prosecutions failed to target the most influential people in the corporate structure.

Foster's study showed that there have been virtually no prosecutions of directors of large companies. It is not hard to see why. It is almost impossible to establish that top people in large corporations are at fault when things go wrong at a particular site. Directors may reside in a distant metropolis and are probably unfamiliar with the technical details of the operations that are under their control. Chief executive officers and company directors can be expected to be diligent about setting up a safety management system and ensuring to the best of their abilities that it is working but, assuming that they have carried out these obligations, they can hardly be held to be personally at fault when things go wrong at a particular site.

There are now suggestions that regulators should develop codes of practice which would make it possible to prosecute company directors who failed to comply.[1] However, we can reasonably assume that, if and when this happens, directors of most large companies will be careful to ensure that they are in compliance. The result will be that, even though a large company may be found to have breached the law, directors will remain immune from prosecution.

Beyond due diligence

Even though there may be no personal fault at the highest level, there is still good reason to find ways to target people at this level. The problem is that, having exercised due diligence and complied with any relevant code of practice, top corporate personnel may view themselves as having discharged their legal obligations and thus feel that they are entitled to rest on their laurels. After all, if safety is the first item on the board agenda, if directors regularly scrutinise audit reports and safety statistics, if they require and receive assurances from senior managers that the company is in compliance with all relevant standards, what more can reasonably be asked of them?

The fact is, we can and must ask more — as was concluded in a review of safety in NSW coal mines. The review found that, despite the best intentions of the people at the top, there was what it called a "disconnect" between the policies developed at corporate headquarters and the practices occurring at the coalface. Its comments are worth quoting:

> "[T]here is a *disconnect* between the intentions of both the DPI [the inspectorate] and the companies, on the one hand, to reduce risk through systems and management plans and, on the other, the reality of risk encountered at the 'coal face' ... [T]he Review stresses the importance of effectively checking (monitoring, observing, inspecting and auditing), so as

---

1   See McCallum, R, Hall, P, Hatcher, A and Searle, A, *Advice in relation to workplace death, occupational health and safety legislation & other matters*, Report to WorkCover Authority of NSW, 2004. McCallum recommended "a code of practice that identifies the following individual responsibilities of management personnel:

   1. the obligations and requirements for an adequate safety management system;

   2. the development of a relevant risk assessment plan; and

   3. the methodology and strategies required and that should be followed for implementing, overseeing and enforcing compliance with relevant safety management systems".

   (NSW WorkCover, *Review of the Occupational Health and Safety Act 2000: discussion paper*, June 2005, p 31.)

to ensure that risk-based management systems and plans are not only in place, but are actually implemented. The Review emphasises that a risk-based management system/plan that is not adequately implemented may be more dangerous than having no system/plan at all."[2]

This is not the first time that this problem has been noted. The inquiry into the Esso Longford gas plant explosion found that the safety management system was "divorced from operations in the field" and "diverted attention from what was actually happening" (Hopkins, 2000:84).

If a company has failed to ensure the safety of its employees, notwithstanding the fact that its most senior people may have exercised due diligence, we can reasonably ask these people to try even harder and to be even more attentive to what may be going on at the operational level of their organisation. Directors need to be sensitive to the fact that bad news travels slowly, if at all, up corporate hierarchies, and they must be willing to bypass normal reporting lines and develop more direct methods of discovering what is happening at the grass roots level (Hopkins, 2005:89, 90).

The Chief of the Royal Australian Air Force recently demonstrated such an approach.[3] He had received a complaint from a lowly corporal that maintenance activities were not what they should be. He thereupon commissioned a special team to visit air force bases and report to him on how extensive the problem was. In this way, he was able to bypass the normal air force chain of command and inform himself far more directly about what was happening.

Another strategy is for directors to walk around regularly at work sites, asking employees to tell them about the problems that they face (Hopkins, 2005:10). This can elicit vital information which might not otherwise be available, including examples of routine non-compliance.[4]

These kinds of interventions on the part of directors go beyond most conceivable due diligence requirements, yet they are the kinds of interventions that are necessary if the disconnect described above is to be rectified. The question to be addressed here is: can the law be used to achieve this end?

Responsibility without fault

It is clear that, if we are to impose some kind of legal responsibility on top corporate people (whom we assume have exercised due diligence), it needs to be responsibility without fault. The law generally requires fault before liability can be imposed (but see the discussion below), so the idea that we might hold top corporate officers responsible in the absence of fault is clearly one that needs careful justification.

---

2   Wran, N and McClelland, J, *NSW mine safety review*, Report to The Hon Kerry Hickey MP, Minister for Mineral Resources, February 2005, pp 7, 8.

3   Author's fieldwork.

4   Author's fieldwork.

The work of legal philosopher Honoré provides a way forward. The starting point for his discussion is the meaning of responsibility in everyday life. He contends that in everyday life we are responsible for the outcomes of our actions, regardless of fault. He terms this "outcome responsibility". Honoré provides the following example. If I trip someone quite by accident, I incur a moral obligation. "An apology is called for, and the person who has been tripped must be helped up and if necessary taken for treatment" (Honoré, 1999:127). Here is another example that Honoré endorses.[5] I am at a dinner party with other guests who are not well known to me. We are discussing a news item about a woman who was raped but failed to report the matter to the police. I express the view that she ought to have made a report. It turns out that one of the dinner party guests was raped earlier in life and chose not to report it to the police. She is deeply upset by my comment and has to retire from the table.

This outcome is hardly my fault. In an important sense, it is just bad luck. Merely by acting in the world, we run the risk of unlucky outcomes of this nature. Nevertheless, I am responsible for the outcome and should do all I can to make amends. I will have incurred discredit by my comment.

Outcome responsibility can also be found in more structured social settings, such as games, professional codes of practice, and traditional codes of honour (Gardner, 2001:131). In the game of soccer, for instance, a free kick can be awarded against a player whose hand touches the ball quite by accident.

Outcome responsibility is also quite explicit in Christian Orthodox liturgy, where God is asked to forgive sins that are "voluntary and involuntary", "known and unknown".[6] According to Orthodox theology, a person may not be at fault for sins that are involuntary, unconscious and unknown, but he or she must nevertheless accept responsibility for them and ask forgiveness.

Honoré develops what he means by outcome responsibility in the following passage:

> "Outcome responsibility means being responsible for the good and harm we bring about by what we do. By allocating credit for the good outcomes of actions and discredit for the bad ones, society imposes outcome responsibility; though often the rewards it attaches and, outside the law, the sanctions it imposes are informal and vague. Under a system of outcome responsibility we are forced, if we want to keep our social account in balance, to make what amounts to a series of bets on our choices and their outcomes. Provided we have a minimum capacity for choosing and

---

5   Honoré, T, personal communication.

6   See, for instance, the communion prayers.

acting, we win the bets and get credit for good outcomes more than we lose them and incur discredit or bad ones. We have to take the risk of harmful outcomes that may be sheer bad luck and not our fault; but that does not make the system unfair to people who are likely to be winners overall" (Honoré, 1999:15).

Honoré argues that outcome responsibility is an inevitable and vital part of personal identity. It is so because identity is bound up with biography; in a sense, we are what we have done (Honoré, 1999:128). If I set out to do something and I succeed, I normally expect to take the credit, even though my achievement depends partly on other people and on luck (Honoré, 1999:130). Moreover, I become identified with that achievement in my own eyes as well as the eyes of other people. To take a pertinent example: if I have managed a coal mine successfully, I become a successful coal mine manager, even though that success may have a lot to do with favourable market conditions. If I take the credit for successful outcomes, I must also take the discredit for unsuccessful or harmful ones — even though these outcomes result partly from the actions of other people and from bad luck. Just as the successful outcomes become part of my biography and my identity, so must the unsuccessful outcomes. If we were not assigned responsibility for outcomes in this way, and if we were not regarded as the authors of the outcomes of our actions, then taking action of any sort in the world would no longer be meaningful. In these circumstances, "we could have no continuing history or character ... [and we] would hardly be people" (Honoré, 1999:29).

Outcome responsibility is not only essential for personal identity, it makes for a better society. It does so by providing "an incentive to aim at and succeed in doing things that are regarded as valuable" (Honoré, 1999:131), as well providing an incentive to avoid or rectify undesirable outcomes. If I am held responsible for unintentionally upsetting the dinner guest, I am motivated to minimise the harm by apologising and I am likely to be more careful at future dinner parties.

Honoré argues that outcome responsibility is the basic type of responsibility in a community — more fundamental than moral or legal responsibility (Honoré, 1999:27). A famous English case illustrates this claim.[7] A batsman smashed a ball out of the cricket ground and onto a highway where it hit the plaintiff on the head. She sued the cricket club for negligence but the House of Lords, on appeal, found that the outcome was so unlikely that the defendant was not at fault for failure to foresee and forestall the injury. The defendant was therefore under no legal obligation to pay compensation. The judges were clearly uncomfortable with this result. One said that the cricket club could fairly be expected to pay compensation, but the law of negligence was concerned with fault not fairness.

---

7   *Bolton v Stone* [1951] AC 850. See Epstein (1973:169, 170).

The English press was outraged by the decision and the cricket club assured the public that, despite the finding in its favour, it would allow the injured woman to retain the damages and the costs awarded by the lower court. In this matter, then, outcome responsibility trumped fault-based responsibility.[8]

Having established the priority of outcome responsibility in everyday life, Honoré argues that "the main role of legal liability is to reinforce our basic outcome responsibility with formal sanctions such as compensation or punishment" (Honoré, 1999:15). The criminal law often adds a requirement of fault before offenders can be severely punished, and damages are generally only awarded if the author of the damage was at fault in some way. But there are many cases in law where the legal response is determined by outcome, in addition to fault. For example, murder is dealt with more severely than attempted murder, even though the intention is the same in both cases, and it may be nothing but luck that determines whether the attempt is successful. Similarly, causing death by dangerous driving is punished more severely than dangerous driving, even though the level of fault is the same in both cases. It is the outcome that makes the difference, despite the fact that luck clearly plays a big part in whether or not any particular instance of dangerous driving culminates in a fatality. Some have argued that holding people absolutely liable for outcomes in this way is unjustified as it is contrary to normal principles of criminal law, but Honoré takes the view that such laws reflect wider social attitudes and that we are right to judge a crime by its outcome, not just by the degree of fault (Honoré, 1999:31).

There is another circumstance, much closer to the concerns of this book, in which the law holds people responsible for outcomes, regardless of fault (that is, it imposes absolute liability). A famous 19th century case, *Rylands v Fletcher*, involved a reservoir that overflowed to neighbouring property, causing damage. The overflow was in no way the fault of the owner of the reservoir, who was nevertheless held liable for the damage.[9] The case established an important precedent. According to Honoré, the law nowadays generally imposes civil liability, regardless of fault, "on those who pursue permissible but dangerous

---

8 A similar case occurred recently in NSW *Coca-Coca Amatil (NSW) Pty Ltd v Pareezer & Ors* [2006] NSWCA 45. A man who had a contract to refill Coca-Cola vending machines at various locations was shot and severely injured in a robbery attempt. The Supreme Court ordered Coca-Cola to pay $3m in compensation, but the Court of Appeal ordered the man to pay the money back, finding that Coca-Cola was not legally liable for his injuries. The NSW Premier asked the company to show compassion for the man and the union said that Coca-Cola had "chosen to refuse to accept accountability for this workplace incident and had left [the man] and his family to fend for themselves without even paying his medical expenses". Clearly, the public perception was that Coca-Cola was responsible for the outcome of the situation it had set up, regardless of fault (AAP, 17/3/06). The company responded to this perception and agreed to compensate the man an undisclosed amount, even though it was under no legal obligation to do so. It said that it "regrets the pain and suffering [the man] and his family have been through and are pleased that he and his family will be able to face the future in a positive way" (AAP, 23/3/06).

9 *Rylands v Fletcher* (1866) LR 1 Ex 265. See Friedman (1972:163) and Gardner (2001:123). The case is no longer authoritative in Australian law but the principle remains unchanged.

activities: storing explosives, running nuclear power stations, keeping wild animals, [and] marketing drugs or other dangerous products" (Honoré, 1999:23, 27).

Clearly, running a coal mine falls into the category of permissible but dangerous activities.

Honoré argues that the law is morally justified in imposing outcome responsibility in such cases. The people who carry out these dangerous activities normally benefit from them, often handsomely, and fairness requires that they accept responsibility for the consequences when things go wrong (Honoré, 1999:24). Gardner (2001:125, 133) adds an additional argument in his commentary on Honoré. Provided potential defendants are warned that their activity is one which attracts absolute liability, there is no unfairness. They can avoid liability simply by not engaging in the hazardous activity.[10]

Honoré's ideas about grounding responsibility in outcomes rather than in fault seem, at first sight, surprising. His real value in the present context is that he shows that this approach is not at all esoteric, but originates in everyday life. What this means is that, insofar as the law imposes outcome responsibility, this should not be seen as an aberration from general legal principle, but an expression of a principle more fundamental than any other. His ideas therefore provide a basis for the project of this chapter, which is to find a way to impose responsibility without fault on the most senior corporate officers.

Before focusing in on this task, however, it is worth considering certain other regimes where outcome responsibility has been practised, or at least discussed — regimes that might conceivably provide a model for imposing outcome responsibility on the most senior corporate officers. Two examples will be considered: the Westminster system of ministerial responsibility, and the Japanese custom of holding top corporate officers responsible.

Ministerial responsibility

Ministerial responsibility has a variety of meanings in the Westminster system. The relevant meaning here is the idea that a minister is responsible for the actions and policies of his or her department. But what precisely does this mean? According to one view, "the act of every civil servant is *by convention* regarded as the act of his Minister" (Finer, 1956, emphasis added). That being the case, any

---

10 Gardner's argument applies when the activity is aimed primarily at private benefit. Braithwaite (personal communication) raises the issue of dangerous activities that have great social value, such as peacekeeping operations in war-torn countries. Peacekeepers might be forced to kill innocent people. Individual peacekeepers cannot choose to withdraw from this situation, but a government can. The fact that it chooses not to should not be taken as *increasing* its degree of responsibility for the undesired outcomes. Nevertheless, if peacekeepers are forced to kill innocent people, we would surely want governments to accept some responsibility and make whatever amends are possible. Honoré's ideas have been exhaustively evaluated in a festschrift; see Cane and Gardner (2001). See also Simons (1997).

serious error by a civil servant must be regarded as the minister's error, or at least an error for which the minister must take personal responsibility. If the matter is serious enough, the minister should resign from office. This has been described as the classic version of ministerial responsibility (Kam, 2000:366). It requires a minister to accept responsibility, regardless of whether the minister knew or could have been expected to know about the matter in question. In Honoré's terms, this is outcome responsibility.

Modern forms of ministerial responsibility are less absolute, requiring ministers to accept responsibility only if they knew or should have known about the matter. Thus, for example, the code of ministerial responsibility promulgated by the Australian Prime Minister, John Howard, states that ministerial responsibility:

> ". . . does not mean that ministers bear individual liability for all actions of their departments. Where they neither knew, nor should have known, about matters of departmental administration which come under scrutiny, it is not unreasonable to expect that the secretary or some other senior officer will take the responsibility."[11]

There is considerable scepticism about whether the so-called convention of strict ministerial responsibility ever operated in practice. One of the most exhaustive studies of this matter was carried out by Finer in 1955 in the UK, and it covered ministerial resignations in the preceding 100 years. Finer was interested in resignations which had been required by parliament or the prime minister or which had been forced by parliamentary criticism. He found only about 20 cases. Of these, 10 were cases where the minister was clearly not personally involved in the matter, but was held vicariously responsible for the failures within his or her department. Finer comments that this is "a tiny number compared with the known instances of mismanagement and blunderings" (Finer, 1956:386). He concludes that there has never really been a convention that holds ministers absolutely responsible for the misdeeds of their departments, and the small number who have actually been held to account in this way were "plain unlucky" (Finer, 1956:394).

In recent years, there has been an increasing tendency for ministers to avoid responsibility by blaming senior civil servants (Kam, 2000:366). For example, an inquiry into the illegal jailing of an Australian citizen by the Australian Department of Immigration concluded that it was the result of a departmental culture that had persisted under successive ministers. The report stated that "there are serious problems with the handling of immigration detention cases. They stem from deep-seated cultural and attitudinal problems within [the

---

11   Prime Minister John Howard, *A guide on key elements of ministerial responsibility*, Canberra, December 1998, p 13.

department] and a failure of executive leadership".[12] It can reasonably be argued that ministers should accept responsibility for the culture of the organisations they head,[13] but the government did not hold the minister accountable in this way. Its response was to move the departmental head. Clearly, we are now further away than ever from strict ministerial responsibility.

Nevertheless, the arguments in favour of ministerial outcome responsibility remain strong. Using a games theory approach, Kam has recently shown that when ministers are governed by a doctrine of strict ministerial responsibility, they will police their departments more effectively (Kam, 2000). They are motivated to search out bureaucratic errors and they have an incentive to correct bureaucratic drift, that is, drift away from officially required practice.

It is not necessary that every case of serious departmental error be followed by ministerial resignation for the incentive to operate. As long as some such cases occur and as long as ministers are aware of this possibility, the incentive exists. It is therefore possible that, in the period covered by Finer's study, the threat of being held strictly accountable for departmental errors did have some of the benefits that Kam describes.

There are several features of this discussion that are relevant to the accountability of top corporate officers. The first is that punishments do not need to be those imposed under conventional criminal law in order to provide the necessary incentives. It is the shame of forced resignation that provides the incentive to seek out departmental error. If outcome responsibility is to be imposed on top corporate officers, it may be more useful to impose shame-inducing consequences rather than conventional retribution. The second is that imposing outcome responsibility induces far more vigorous efforts than is the case when responsibility is fault-based. It obliges senior officers to search out and correct bureaucratic drift (to use Kam's term) — and they can never be certain that they have fully discharged this obligation. On the other hand, where responsibility is fault-based, lawyers can advise senior officers on what they need to do to discharge their obligations, after which they need do no more. Finally, Kam's reference to bureaucratic drift is particularly relevant to large corporations. Snook has argued that it is the key to understanding what goes wrong in the hyper-complex organisations of the modern world (Snook, 2000:235). Corporate officers need to be as alert as possible to the ways in which practice may drift away from policy. This is precisely the issue of "disconnect" referred to earlier.

---

12 Palmer, M, *Inquiry into the circumstances of the immigration detention of Cornelia Rau: report*, Canberra, Commonwealth of Australia, July 2005, p xi.

13 Macintosh, A, "The buck no longer stops with this government", *The Canberra Times*, 19 July 2005.

The responsibility of Japanese corporate heads

The Japanese corporate world is a well-known environment in which outcome responsibility is attributed to the head of the organisation. If some event incurs shame for the corporation, Japanese custom requires the head of the corporation to take personal responsibility, apologise and, if necessary, resign — even though he or she may not be personally at fault in any way.

A recent example of this is the resignation of the President of the Tokyo Stock Exchange in December 2005, along with the head of the exchange's computer system and the exchange's managing director. A trader, acting on behalf of a bank, had mistakenly placed a huge order to sell. When the trader discovered his mistake, he tried to have the order cancelled but a fault in the stock exchange computer prevented him from doing so for about 10 minutes. In the meantime, other traders snapped up the shares at a bargain price, costing the bank US$347 million. Other major stock exchanges would have picked up the error immediately and suspended trading. According to one commentator, what was needed was "a full redesign of the bourse's trading system, which is often described as the most complicated in the world. The problem is that it is essentially based on procedures used when transactions were processed by hand".[14]

The President had been in his job for less than two years and was not personally at fault for the antiquated state of the exchange's trading system. Yet he took full responsibility by resigning. As one newspaper reported, he "stood in the glare of television cameras . . . to do what Japanese executives perhaps dread most: bow in shame".[15]

Fisse and Braithwaite describe this situation as one of *noblesse oblige* (literally, nobility obliges), which "means the titular head of the organisation assuming strict individual responsibility for collective wrongdoing" (Fisse and Braithwaite, 1993:112). This is a ubiquitous and long-standing feature of Japanese culture, applying to business leaders and government officials of all types. According to one observer of Japanese ways, "official responsibility means the assumption by the public officer of the consequences, however remote, of his official acts" (Duran, 1937:4). Unlike the Westminster convention of individual ministerial responsibility, Japanese *noblesse oblige* operates in practice. The suicide by Japanese military commanders whose forces fail in battle is a well-known example. Even in times of peace, Japanese military commanders may commit suicide because of mistakes made under their command for which they have been personally exonerated (Duran, 1937:12).

---

14   Editorial, "TSE chief needs expertise, leadership ability", *Nikkei*, 22 December 2005.

15   Pesek, W, "The blame game in Tokyo", *Bloomberg News*, 25 December 2005.

The willingness of Japanese business leaders to take personal responsibility in the absence of personal fault must be seen in the context of Japanese company decision-making. So-called *ringi* management involves the formulation of draft plans at the middle management level, followed by exhaustive consultation both horizontally (by other groups of equivalent rank), as well as vertically (by people at other levels in the hierarchy). By the time the plan is ready for approval at the highest level, it is everybody's and nobody's. At this stage, as one observer has said, "the nature of the act of confirmation or approval is very vague . . . this is a skilful device for obscuring both the locus of authority for carrying out the plan and the locus of responsibility for its effects" (Brown, 1974:182).

According to another observer, "the Japanese . . . have resisted the urge to parcel out a firm's myriad operations in neat little bundles of authority and responsibility" (Brown, 1974:187).

This situation means that failures cannot be meaningfully attributed to the actions or inactions of particular individuals. As a result, no one, and certainly not the CEO, is personally at fault, in any western sense, when things go wrong. This has led some observers to claim that the apologies and resignations of Japanese leaders are merely symbolic or ritualistic, and that the Japanese system involves the "symbolic sacrifice" (Fisse and Braithwaite, 1993:112) or "scapegoating" (Dore, 1973) of its leaders. Fisse and Braithwaite are particularly critical:

> "Why should an individual person be sacrificed rather than the corporate ox that gored? For a collectivist culture, it is perverse that Japanese law does not direct more of the fire and brimstone of public shame at corporate entities rather than individuals" (Fisse and Braithwaite, 1993:112).

Whether or not it is perverse, there is certainly a puzzle here. Western societies do not insist on finding someone to blame when things go wrong. Criminal law is on the look out for individual fault but, where none exists, it can still hold corporations to account. Why does the Japanese system require what to western eyes appears to be a ritual sacrifice?

The answer, surely, is related to the way in which the "self" is constituted in Japanese society. Japanese personal identity is bound up with group identity to a greater extent than it is in the West. This is true, in particular, for Japanese business leaders and their companies. Given this identification of the individual with the company, the collective fault of the organisation is simultaneously the fault of the head of the organisation. The corporate leader *feels* personally responsible for the collective failure, and it is precisely for this reason that a public apology is a deeply felt and personally shameful experience, not merely a ritual.

The fact that a Japanese leader might feel personally ashamed in these circumstances seems, at first sight, strange to western observers. Sociologists have shown, however, that personal identity is bound up with group identity — even in societies which purport to be far more individualistic. This is most easily seen in the case of national identity. As an Australian travelling abroad, I have felt personally ashamed of some of the actions of my government. In the eyes of others I meet and even in my own eyes, I am diminished by those actions, simply because I am an Australian. Of course, I am not personally implicated and may even seek to distance myself from those actions by explaining that I voted against the government, but I cannot disown my national identity and so cannot escape some degree of responsibility.

In times of war, the identity of the individual is tied so tightly to the nation that individuals may have to answer with their lives for the actions of the nation. The taking and killing of innocent hostages in the conflicts in the Middle East is a contemporary example.

The point can be put even more sharply. Although it seems strange to western observers that Japanese business leaders should accept personal responsibility for the faults of their company, that is precisely what westerners expect of Japanese political leaders in relation to the faults of their nation. An Australian prime minister in the 1990s demanded that Japanese national leaders of the day should apologise for Japan's role in the Second World War, 50 years earlier. The expectation was that present-day Japanese political leaders, who were no more than children and may not even have been alive during the Second World War, should express personal remorse for Japan's role in that war. At this point, the thoughtful reader might suggest that what was expected was not a personal apology but an apology *on behalf of* the nation. But apologies are only useful if they are sincere, that is, accompanied by genuine remorse and shame. So it was that, on the fiftieth anniversary of the end of the war, the Japanese Prime Minister issued a statement which said: "I ... express here once again my feelings of deep remorse and state my heartfelt apology." The Australian Prime Minister accepted this apology. In expecting a Japanese political leader to experience shame, we are assuming that his personal identity is in some way bound up with that of his nation and that the faults of his nation are to some extent his own.[16]

These comments are intended to demonstrate that *noblesse oblige*, Japanese style, is not as foreign to western ways of thinking as might at first appear. This is not some curious phenomenon of anthropological interest only. It is a way of thinking that has parallels in western societies and has, moreover, deep sociological roots.

---

16   *The Sun Herald*, 27/5/95, 19/8/95; *The Sunday Age*, 12/8/95.

The obligation of Japanese business leaders to accept responsibility for the faults of their companies is customary only. It is not a component of Japanese criminal law. Fisse and Braithwaite argue that titular responsibility of this type cannot be part of western criminal law either, because it is fundamentally unfair. What we can do, they say, is impose criminal liability on corporations and then hope that "organisations imbued with the ethos of *noblesse oblige* will translate this quickly enough into assumptions of responsibility by the titular head" (Fisse and Braithwaite, 1993:113).

The problem is that western companies do not often function in this way. Corporate leaders will not take personal responsibility for deaths that happen at remote company outposts, unless the law finds ways of imposing it on them. Some of the most safety-conscious multinationals summon plant or site managers to corporate headquarters to account for deaths at sites which are under their control and to explain what they intend to do to ensure that no further fatalities occur. This is an internal system for imposing outcome responsibility on site managers. Such a practice is likely to have a beneficial effect on site safety and is therefore to be applauded. But it does not impose responsibility on corporate leaders. To repeat, while Japanese culture imposes outcome responsibility on corporate leaders, there is nothing in western culture or western law that imposes outcome responsibility on top corporate leaders, in the absence of personal fault. The challenge, then, is to find a legal mechanism to do for western companies what the national culture does for Japanese companies.

Imposing outcome responsibility on corporate leaders

Let us recall some of the arguments made earlier in this chapter. First, corporate leaders are often immune to fault-based liability since, in many large corporations, they can be expected to be complying with due diligence requirements, even when things go wrong. Holding them strictly responsible for outcomes is one way of going beyond due diligence and establishing legal incentives to attend more closely to safety. Second, outcome responsibility is not some legal aberration; it is the most basic form of responsibility. Honoré shows, moreover, that outcome liability is morally justified. Where organisations are engaged in dangerous activities — as many large corporations are — and where their leaders reap the benefits of those activities, it is only fair that they also shoulder responsibility for any harm caused by those activities. Particularly where they are warned about their liability, it cannot be said that outcome responsibility is unfair.

Assuming that senior company officers are to be held legally responsible in this way, what kinds of sanctions should be imposed? The courts can hardly force them to resign, Japanese style, but they could prohibit them from acting as directors or CEOs for a number of years. Would this be an appropriate response? Here, I turn again to Honoré. In an important passage, he notes that, while "it is a myth that fault and desert are essential to responsibility", this is not to say that they are irrelevant. "They serve rather to increase the credit or discredit for the outcome of our behaviour that we incur in any event" (Honoré, 1999:31). In the Japanese collectivist tradition, the fault of the corporation shifts subtly to the individual leader. Typically, that does not occur in a more individualist culture. Indeed, the situation contemplated here is one where the leader is not at fault in any significant way. In these circumstances, the discredit due to the leader is limited and it would be inappropriate to remove him or her from office.

On the other hand, to impose a nominal fine would be to miss a golden opportunity. There is now a strong tradition in criminology that a purely punitive response is not the best way to achieve the objectives of criminal law. Punishment in and of itself does not create remorse in the offender and therefore provides no moral incentives to do better. Moreover, the levels of punishment that can be justified by fault are often far less than the losses and suffering experienced by victims and leave victims feeling unacknowledged and hence unable to move on in their lives. The restorative justice movement in criminology seeks to overcome these defects. Perhaps not coincidentally, one of the sources of inspiration for this movement is the Japanese response to crime in general and to corporate crime in particular. The aim of the system is to shame the offender, induce remorse in him or her, and secure a sincere apology for the victim which, precisely because it is sincere, is likely to involve some form of material restitution in addition to its symbolism.

Braithwaite and others have argued that, because neither national cultures nor conventional legal systems in the West achieve these outcomes, we need to create institutions to fill the vacuum (Braithwaite, 1989). One such institution is the community conference in which, following a plea of guilty, the offender meets with the victim or victims, together with other people who, while respected by the offender, nevertheless do not condone the offender's actions. The meeting is designed to achieve various ends:

- to confront the offender with the full consequences of the offence
- to shame the offender — not so as to ostracise, but in a way that induces remorse and a commitment not to reoffend
- to secure recognition for the victim of the suffering inflicted
- to secure a sincere apology for the victim, and
- to impose some practical consequences on the offender, for instance, payment of financial compensation or attendance at some relevant course (Braithwaite, 2002:24-26).

Fisse and Braithwaite provide a powerful example of how this can work in the context of corporate crime. Agents for several insurance companies had been selling worthless life insurance policies to Aboriginal people living in isolated parts of Northern Queensland. The agents had pressured Aboriginals in various ways — even threatening that they would be imprisoned if they didn't sign up.

Saddest of all, the Aboriginals were falsely told that the policies would pay generous funeral benefits that would help to transport bodies back to the place of origin for burial — a matter of profound importance to these Aboriginal people. When these matters finally came to court, senior corporate officers denied knowledge of the practices. Nevertheless:

> "... for some participants, responsibility was brought home in a particularly compelling way. Top management found themselves directly confronted with the shame of the practices from which they and their companies had benefited. The media and the courts were not the only forums in which some found themselves exposed. Top management from Norwich (one of the companies concerned) were pressed into immediate contact with the victims as part of the process leading up to the settlement. This was an exacting and conscience-searing experience. They had to take four-wheel-drive-vehicles into Wujul Wujul (in tropical North Queensland) to participate in dispute negotiations in which the victims were given an active voice. Living for several days under the same conditions as their victims, Norwich's top brass had to sleep on a mattress on a concrete floor, eat tinned food, and survive at times without electricity" (Fisse and Braithwaite, 1993:236).

It is worth quoting Fisse and Braithwaite further on the consequences of this kind of experience:

> "Processes of dialogue with those who suffer from acts of irresponsibility are among the most effective ways of bringing home to us as human beings our obligation to take responsibility for our deeds. Traditional courts, where victims are treated as evidentiary cannon fodder rather than given their voice, have tended to be destructive of this way of eliciting responsibility ... Boardrooms and executive suites are hardly the frontiers where victims are harmed, but provide a haven conducive to cosy rationalisations and distorted pictures of actual corporate impacts ... Encountering victims allows the shame of the wrongdoing to be communicated directly to those responsible. The process of encounter also helps to pre-empt or counter efforts by directors or managers to deny the existence of the problem or to neutralise it by means of some self-serving rationalisation. Beyond these salutary effects, encounters with victims provide an opportunity for healing through acceptance of responsibility and putting right the wrong" (Fisse and Braithwaite, 1993:236).

This passage is important for more than one reason. It clearly describes the cathartic effects of meetings between top corporate leaders and victims of corporate wrongdoing. But, beyond this, it is interesting for what it says or implies about the responsibility of top corporate leaders. It talks about "our obligation to take responsibility for our deeds", about "eliciting responsibility", and about encouraging the "acceptance of responsibility". This is not a fault-based conception of responsibility. The corporations were at fault, as were a number of their agents. But Fisse and Braithwaite are not talking about the transfer of fault from the corporation to its senior executives, Japanese style. They are talking about how meetings with victims encourage senior executives to accept personal responsibility for outcomes, regardless of personal fault.[17] This is precisely Honoré's idea of outcome responsibility.

Senior officers who accept responsibility for outcomes in this way will act decisively to correct the problems that led to the offence. They will not be satisfied with the practices of due diligence, which too easily degenerate into rituals to protect senior officers from legal liability. They will be motivated to ferret out errors and wrongdoing — not just to ensure that they are personally immune from their consequences.

The consequences for the senior executives described in the life insurance case are precisely the types of consequences that the law needs to impose on senior officers of large corporations when people are killed. Managers who have been in charge of a mine at which a death occurs will sometimes tell you that it is a life-changing experience. They usually know the person killed, sometimes quite well; they are acutely aware of the impact on fellow workers, who may need counselling; and they may have had the unforgettable task of conveying the sad news to the family. What is needed is a legal mechanism for imposing these consequences on senior corporate managers. Chapter 2 described the sense of frustration felt by relatives and friends of the dead Gretley miners at their inability to get senior company people to accept responsibility. Legally imposed outcome responsibility would satisfy this community expectation.

A legal model

I hesitate to propose the precise wording for an amendment to the OHS legislation that would achieve this end, since critics will no doubt focus on the problems with this particular formulation, without considering the more fundamental issues. But, to be concrete about this, let me suggest a possible model, using the NSW OHS Act as starting point. Section 26 of the Act states that:

> (1) If a corporation contravenes . . . any provision of this Act . . . each director of the corporation, and each person concerned in the management of the corporation, is taken to have contravened the same provision unless the director or person satisfies the court that:
> (a) he or she was not in a position to influence the conduct of the corporation in relation to its contravention . . . , or
> (b) he or she, being in such a position, used all due diligence to prevent the contravention . . .

---

17   Fisse and Braithwaite conceptualise this a little differently. They argue that the fault or otherwise of senior executives depends on how they *react* to corporate offences. They describe this as reactive fault. According to this formulation, where senior executives react in a non-restorative fashion, they are personally at fault (1993: 210-213).

Conviction under this section involves an inference of fault, that is, that the person did not exercise due diligence. An additional section imposing outcome responsibility might read as follows:

> (1) If a corporation contravenes any provision of this Act, and
>
> (2) if, as a result of that contravention, a person is killed,
>
> each director of the corporation, and each person concerned in the management of the corporation, is guilty of an offence,
>
> unless that person satisfies the court that he or she was not in a position to influence the conduct of the corporation in relation to its contravention.

There are a several things to note about this formulation. First, unlike section 26, it only comes into operation if the corporate violation has resulted in harm (in this case, death). In these circumstances, it requires senior officers to accept responsibility for the outcome. Second, it therefore falls short of absolute outcome responsibility. It only comes into operation if death is a result of an offence committed by the corporation. This is a compromise, but it brings the situation into line with the life insurance case (described earlier), in which outcome responsibility was imposed on top corporate officers only after corporate fault had been established. Third, there is no liability if a defendant can establish that the violation and death were not, in fact, outcomes in some sense of the defendant's actions or inactions. Fourth, as with section 26, it targets a potentially large number of senior corporate officers. It would be up to the prosecution to decide which individuals could be most usefully charged but, unlike section 26, the prosecution would be able to target people at the apex of the corporation.

The penalty under this provision might be to attend a community conference with relatives of the victim to hear their grief and to explain what the corporation intends to do to rectify the wrong and to prevent any recurrence. One goal of such a conference might be for the corporate officer concerned to express an apology that relatives of the victim agree to accept. Relatives might impose a number of conditions on acceptance, for example, the apology must be written and published, it must be accompanied by various promises, and so on. Obviously, the processes of the conference would need to be described in detail, perhaps in a code of some sort. There would need to be backup provisions in the event that a senior officer refused to engage in the process or refused to engage in good faith. It might be, for example, that the company itself would be subject to heavier penalties in the event of non-cooperation by its senior officers.

I repeat, this specific legislative formulation is not intended to be the last word; it is outlined here simply to provide an indication of what outcome liability might look like in practice.

Finally, it should be stressed that a provision imposing outcome liability would not replace any existing provision imposing fault-based liability. Where a senior officer is at fault, there may be good reasons to make use of the fault-based provision. For one thing, heavy penalties can be imposed, if appropriate. Second, it may be desirable to take legal action where dangerous conditions have been allowed to persist but, simply by luck, no one has been killed or injured. Outcome liability is of no use in these circumstances.

Chapter 8
Organisational mindfulness

Previous chapters have identified two rather different ways in which the Gretley accident might have been averted. The first (on which the mine placed exclusive reliance) was to identify the location of the old workings before work commenced, and to plan mining operations, so as to keep at a safe distance. Management failed in this respect and it was this failure on which the prosecution focused. The second way that the accident might have been averted would have been to respond more diligently to the warnings of danger that began to emerge in the days prior to the accident.

These two strategies represent two rather different approaches to risk management. The first I shall call the standard risk management strategy. Before an operation commences, it requires management to carefully identify all potential hazards and to put controls in place to deal with those hazards. This is the strategy of "getting it right in the first place". The second strategy starts from the assumption that one can never be sure of getting it right in the first place. It is therefore necessary to be alert to indicators of danger that may emerge during ongoing operations and to respond to these in a conscientious way before disaster strikes. These are not, of course, mutually exclusive approaches; they are complementary, constituting, if you like, a first and second line of defence against disaster. This chapter discusses these two approaches in more detail and highlights the importance of the second.[1] Had the Gretley management recognised the importance of this second line of defence, the accident would not have occurred.

The standard risk management strategy is embodied in much contemporary legislation. It is expressed in its most developed form in the safety case regulations that apply to many major hazard facilities in Australia.[2] Safety case regulations explicitly require major hazard facility operators to identify hazards, assess risks and put controls in place to deal with those risks. Furthermore, they require facility operators to establish safety management systems to ensure that these controls remain in place. Once an operator has developed its case in this way, it must present it to the regulator for approval. The regulator scrutinises the case, requests changes if necessary, and ultimately authorises the operation.

There is a strong emphasis in this approach on "getting it right in the first place". Moreover, there is an implicit presumption that, once the licence has been obtained, the ongoing process of safety management becomes more

---

1   The chapter draws on Hopkins (2002a).

2   See, for example, the Occupational Health and Safety (Major Hazard Facilities) Regulations 2000 (Vic).

straightforward. It is certainly not the intention of safety case regulations that operators pay less attention to safety after they have been licensed, but there is a danger that this is how the regulations will be interpreted by those who manage major hazard facilities.

The second strategy — being alert to warnings of danger — is emphasised by so-called high-reliability organisations, that is, organisations that use hazardous technologies but nevertheless function with extremely high levels of safety and reliability (Weick and Sutcliffe, 2001). Research on such enterprises shows that they are extremely sensitive to the possibility that something may be about to go wrong. They exhibit "chronic unease" about the hazards that they deal with (Reason, 1997:37). They seek out information about small failures in the belief that these may be the pre-cursors to catastrophic failure. They have systems to pick up warnings of danger and to analyse these warnings carefully. They recognise that, no matter how conscientiously they have tried to get it right in the first place, they can never be certain that they have succeeded, and they therefore stress the need for ongoing vigilance. In short, they exhibit what the researchers have described as collective or organisational mindfulness (Weick et al, 1999:100).

The initial Gretley inquiry commented specifically on the way in which mine managers had relied on getting it right in the first place and failed to exhibit the kind of risk-awareness and chronic unease that characterise high-reliability organisations. In particular, the inquiry was critical of managers for failing to seriously consider the possibility that something might be amiss in the days immediately before the accident. The words of the report are worth quoting at some length:

> "In several instances, persons in the mine management hierarchy demonstrated, by their answers to questions in the course of the hearing, an attitude of mind which appeared to make assumptions and act on them without questioning whether or not they were valid. Similarly, on several occasions, conclusions appear to have been readily arrived at (eg that no investigations of a particular matter were required) rather than maintaining an open or questioning mind. A tendency towards closure rather than maintaining a questioning mind is an attitude fraught with danger."[3]

The inquiry spelt out what it saw as the appropriate attitude of mind in the circumstances:

> "Something more than a superficial assessment was called for in the circumstances where mining was taking place in the vicinity of old workings known . . . to be full of water. The terms of Mr M's report were startling, and different. They were the observations of an experienced deputy. The panel was known to be the driest in the mine. How long had Mr M observed the considerable seepage at the face? What was the flow

---

3   These words were put by counsel to the inquiry and the judge quoted them approvingly in his report; see I p 627.

rate of the trickle? Had the water reappeared after production ceased? What was the likely source? If the [old] colliery was a possible source, what did that suggest? Might the plan be wrong?"[4]

Unfortunately, as the inquiry observed, "none of these questions was asked or answered".[5]

Institutionalising the second line of defence

The strategy of being alert to signs of danger has now been institutionalised to some extent in coal industry regulatory regimes in Australia. The new approach is not directly attributable to the Gretley experience; it goes back to an explosion two years earlier, in 1994, at the Moura coal mine in Queensland. Eleven miners died on that occasion. In the days prior to the Moura explosion, there had been warnings of danger that had been dismissed, misinterpreted or generally downplayed (Hopkins, 1999). The fatalities could certainly have been avoided had the warnings been attended to conscientiously.

The nature of the warnings at Moura needs to be understood in a little more detail. Once coal is exposed to air, it has a tendency to heat up, slowly, in a process known as spontaneous combustion. Coal that is heating up in this way must be well ventilated; unless it is, its temperature can rise to the ignition point of methane. Then, if methane is present in combustible concentrations, it will explode.

As the coal is heating up, it gives off increasing amounts of carbon monoxide (CO), which thus serve to indicate that a dangerous situation is developing. Moura's gas monitoring system was registering rising levels of CO in the period prior to the explosion, and these were the warnings that were ignored. Moura was the third explosion in Queensland coal mines caused by spontaneous combustion in the space of 20 years. A total of 41 men were killed in these explosions.

Moura, then, was the catalyst for a new regulatory approach that, among other things, sought to encourage companies to attend to warnings of danger. Legislation in Queensland and NSW now requires mines to develop a specific management plan for each major hazard.[6] Certain subordinate regulatory documents specify that major hazard management plans should identify trigger events, that is, indicators of danger that will necessitate certain specified actions.[7]

---

4 I p 628.

5 I p 628. See also I p 643.

6 *Coal Mining Safety and Health Act 1999* (Qld); *Coal Mine Health and Safety Act 2002* (NSW) (not technically in force at the time of writing).

7 Queensland Government Department of Mines and Energy — Safety and Health Division, *Approved standard for mine safety management plan*, Revision No 3, January 1998, pp 12, 22-24; Coal Mines (Underground) Regulations 1999 (NSW), reg 133 and 134; NSW Department of Mineral Resources, *Spontaneous combustion management code*, Revision No 3.3, 26 August 1996.

For some hazards, there could be several trigger levels (of increasing seriousness), with corresponding action plans, ranging up to withdrawal of all personnel from the mine. Mines have therefore developed schedules of triggers and corresponding actions, and these have become known as TARPs (trigger action response plans).

An example: managing spontaneous combustion

The potential of this new regulatory system is well illustrated by the experience of one particular Queensland mine when dealing with the spontaneous combustion hazard. In response to the regulatory requirements, it set three trigger levels for CO concentrations in areas where mining has been completed and which were sealed off (except for gas monitoring points). They were as follows: less than 50 ppm (parts per million) of CO is normal; 50–120 ppm is a level 1 alarm; between 120 and 500 ppm is a level 2 alarm; and above 500 ppm is a level 3 alarm.

The regulator requires mines to establish incident control groups of predetermined composition to manage such events. Such control groups must contain people of sufficient authority to implement decisions, they must maintain activity logs, and they must not disband until the situation has been resolved.[8] At the mine in question, the composition of the control group varied with the level of alarm. At level 1, the control group consisted of the mine manager alone, while at level 3, management had determined that the control group should contain people from outside the mine, including someone from corporate headquarters, a government inspector and a full-time union safety officer.

A major spontaneous combustion event occurred at this mine soon after it had adopted its plan, and it is worth recounting the details.[9] Carbon monoxide monitoring picked up indications that spontaneous combustion might be occurring in a mined out and closed off area of the mine. The readings were high enough to trigger a level 3 alarm which, among other things, required the withdrawal of men from the mine. The control group was constituted, as described above. After some hours, management came to the conclusion that the readings did not indicate a current problem and that the gas sampling process in the closed off area was picking up remnants from an incident which had occurred five months earlier. The mine manager was concerned about lost production and wanted the men to return to work underground as soon as possible. He put considerable pressure on the control group to endorse his view, but the external members held out against this pressure. An external expert in gas analysis was then called in to review the gas readings, and he concluded that

8   Queensland Government Department of Mines and Energy — Safety and Health Division, *Approved standard for mine safety management plan*, Revision No 3, January 1998, p 24.

9   Details provided by the mines inspectorate.

the situation was urgent, that a fire was burning somewhere underground, and that unless it was extinguished immediately an explosion was highly likely. At this point, management acted decisively, hired an inert gas generator, and pumped the affected part of the mine full of inert gas. Carbon monoxide readings indicated that the fire was extinguished within a matter of hours. The outcome was a vindication of the mine's spontaneous combustion management plan.

In subsequent analysis (Stephan, 2001), the government inspectorate stressed that the experience had highlighted the importance of including outside personnel in incident control groups. There is enormous pressure on control group members to interpret ambiguous indicators in such a way as to allow production to continue. The co-option of group members to management's viewpoint was only prevented by the presence of outside participants — in particular, the union safety official. This was a situation where the well-known phenomenon of "groupthink" (Janis, 1972) might have been expected to operate. One of the techniques by which groupthink can be overcome is if someone is given the role of devil's advocate, with the job of raising doubts about any decision which the group appears likely to make. In the above case, the union official, quite naturally, acted as a devil's advocate and this was crucial in ensuring that the control group functioned optimally.

The spread of the TARP philosophy

The philosophy of specifying triggers and corresponding action response plans (TARPS) is clearly spelt out in regulatory documents in the case of spontaneous combustion. But the principle has been widely adopted by many coal mines in NSW and Queensland, and has been applied to a number of other hazards. It would be fair to say that it is the living heart of many principal hazard management plans. Some examples, drawn from various mines that I have visited, will serve to illustrate the approach.

Take the hazard of a collapsing roof. As mining moves into new areas, the nature of the roof can change; in particular, it can become less secure and need more intensive support in order to protect miners working beneath it. Miners must therefore be alert to the changing nature of the roof. Indicators of increasing danger include falling flakes of rock, increasing quantities of water dropping from the roof, and the appearance of certain geological formations. One mine has identified four states of increasing concern which have been labelled green, yellow, orange and red. Triggers are specified for each state, along with the actions required by miners and their managers. During discussions with miners at this particular mine, I discovered that they were well aware of the trigger levels and the actions required of them.

The flexibility of the TARP approach was demonstrated at an open-cut coal mine that I visited. The huge haulage trucks at such sites constitute a major hazard. If the driver of such a vehicle loses control, it can do enormous damage. If the haulage truck collides with an ordinary vehicle, that vehicle can be crushed beyond recognition. The mine in question had decided that the risk of such an event increased during wet weather. It had therefore defined two triggers: wet roads and severely rain-affected roads, with corresponding restrictions on the drivers of heavy and light vehicles. Obviously, the decision as to whether a road is severely rain-affected or merely wet is not clear cut, and it is the production supervisor's responsibility to exercise this judgment.

Given the subject of this book, a major question of interest must be the possibility of applying the TARP philosophy to the prevention of inrush. Clearly, the indicator must be related to the flow of water at the mine face but, just as clearly, it is difficult to set unambiguous triggers. Here is how one mine has dealt with the problem. Some water, it has decided, is normal. Beyond this, the mine has set three trigger levels: (1) an increase in the amount of water that is seeping out from the mine face and roof and accumulating on the floor; (2) an increase in the *visible flow* of water from the face or roof; and (3) a major change in the flow of water from the face or roof. Although it may be difficult to distinguish these various levels, effective response plans swing into action even at trigger level 1. Supervisors must record observations, and both the mine geotechnical engineer and the mine surveyor must review the situation and think again about potential inrush sources. At trigger level three, mining must cease, personnel must be withdrawn, and mining must not restart until a restart plan has been developed and implemented. Presumably, this might include drilling ahead. It is clear that mines that have developed inrush TARPs are responding directly to one of the lessons that emerged from the Gretley inquiry.

As can be seen clearly in these examples, the TARP philosophy formalises the idea that disasters can be prevented by being alert to warnings and responding in a conscientious fashion. Obviously, some of these TARPs (such as the wet weather and inrush TARPS described above) are somewhat rough and ready, and present difficulties in terms of deciding what trigger state exists. They do not provide a clear-cut set of decision criteria and they do not relieve those in positions of responsibility from the need to exercise judgment about the level of danger that exists. However, they do focus the attention of those in positions of responsibility on the need to make those judgments, and on the need to make them as conscientiously as possible. They structure decision-making by defining the kinds of decisions that need to be made, and they make it much more difficult for managers to ignore warnings — as they have so often done in the past, with disastrous results.

The two lines of defence again

I noted earlier that there were two contrasting strategies for preventing disaster: getting it right in the first place, and ongoing alertness to danger. I need to stress again that these are complementary, not alternative, strategies. They correspond to first and second lines of defence. Regulators emphasise the need for coal mines in Queensland and NSW to have major or principal hazard management plans that control known hazards from the outset. But such plans must also contain ways to ensure that controls remain effective. Trigger action response plans can be seen as one of the ways to ensure that hazards remain under control and, from this point of view, they are part of a single overall strategy. Yet is it clearly useful to identify the TARP as a second and distinct line of defence. The interesting thing is that, while the TARP strategy is not emphasised in the formal safety legislation, mine management has recognised its importance and implemented it in quite creative ways. In so doing, the mining industry has picked up on one of the vital lessons of both the Moura and Gretley mine disasters.

TARPs in high-risk process industries

It is interesting to note that the philosophy of attending to warnings of danger is implicit in the safety case regimes mentioned earlier. The most developed safety case regimes operate in the petro-chemical industries and, overseas, in the nuclear power industry. These are industries that process highly dangerous substances, including toxic chemicals, highly inflammable petroleum products and radioactive materials. If operators lose control, the results can be catastrophic. Safety depends on keeping these dangerous substances safely contained. To this end, the processes must be maintained within certain parameters, that is, limits of pressure, temperature, volume, and so on. If the processes are allowed to stray outside these parameters (limits), the situation can escalate dangerously — sometimes beyond the point of no return. This is what happened in the Longford gas plant accident near Melbourne in 1998, and it is what happened at the BP Texas oil refinery in 2005, when 15 workers were killed. In both cases, without realising it, control room operators allowed certain vessels to fill beyond a safe limit and, in both cases, the vessels overflowed, initiating a disastrous sequence of events.

For this reason, safety case regulations in these industries require facility managers to define "critical operating parameters" (limits of pressure, temperature, etc) as boundaries of operation. If the process temporarily exceeds these parameters, the safety case treats this as a warning sign that the process is not adequately under control and that, although there was no loss of containment, there might have been. Safety case regulations therefore require facilities to record such deviations from normal (safe) operation, count them, and place limits on the numbers of such occurrences. However, they do not always require companies to have a specific action plan to deal with such deviations. To take a leading Australian example, the Victorian major hazard facilities regulations require companies to specify the resources that they have available in response to the failure of a control, particularly a critical control, but they do not require companies to formulate an *action plan* to respond to these deviations.[10] It would be a simple matter to impose on companies the additional requirement that they develop such a plan to deal with situations where critical operating parameters have been exceeded.

This discussion reveals that the spirit of the TARP is not far from the surface in the safety case regulations of major hazard facilities. Exceeding a critical operating parameter is, in effect, a trigger that requires action, and companies are expected to react in some way (although, under the Victorian legislation, they are not required to say just how they intend to react). Safety case regulation is sometimes criticised as placing too much emphasis on the initial phase of getting a safety licence and not enough on the need for ongoing efforts to ensure safety. This criticism could, to some extent, be countered by stressing the features of the safety case that correspond to coal industry TARPs.

Applying the trigger event model more widely

We have seen in the discussion above that the philosophy of identifying triggers and actions to be taken (the TARP philosophy) is in no way confined to coal mining. It is a philosophy that is widely applicable for the prevention of major accidents. Let us consider this wider application in a little more detail.

On the basis of an extensive review of major accident reports, Turner concluded that disasters always involve an information or communication failure of some kind. There is always information available somewhere in the organisation that constitutes a warning which, if heeded, would have averted the accident. For Turner, the central question to ask, then, is "what stops people from acquiring and using the appropriate advance warning information so that large-scale accidents and disasters are prevented?" (Turner, 1978:195). He provides a variety of answers:

- information is noted but not fully appreciated — this may occur for a number of reasons, including a false sense of security which leads individuals to discount danger signs, pressure of work which diverts attention from warning signs, or difficulty in sifting the information from a mass of other irrelevant facts

- prior information may not be correctly assembled — this may be because it is buried in other material, distributed among several organisations, or distributed among different parties in the same organisation, and

10   Occupational Health and Safety (Major Hazard Facilities) Regulations 2000 (Vic), Sch 4.

- bad news is usually unwelcome — as a result, people in possession of relevant warning information may be disinclined to pass it up the line to senior managers.

Based on this analysis, it would seem that the starting point for applying the trigger event model more widely is an effective reporting system. It is important that such a system report not only occurrences with the potential to affect safety, but also hazards, that is, unsafe conditions. It must, in short, be both an occurrence and a hazard reporting system. But, if this is to have any chance of gathering relevant warning signs, management must put considerable thought into specifying what sorts of things should be reported: what are the warning signs that something might be about to go disastrously wrong? Matters which management might decide to treat as warnings include: certain kinds of leaks; certain kinds of alarms; certain maintenance problems, particularly backlogs; machinery in a dangerous condition; corrosion; and procedures that are found to be inappropriate. Management must also ensure that such information is indeed reported or collected. Once identified, these reported events or conditions must be treated as triggers to action, and management must specify what kind of action is required and who is responsible for taking the action. For instance, should a further investigation be undertaken, or should production be stopped? Such a system structures decision-making in a way that forces organisations to take note of and respond to warnings of danger.

Of course, reports need not be restricted to matters that are specified by management. To do so would limit the full potential of a reporting system and discourage the kind of risk-awareness that high-reliability organisations seek to achieve. Accordingly, reporters must be free to report any or all matters about which they are concerned, not just those specified by management. Such reporting systems are now widely used by major airlines and are seen as a vital tool for the maintenance of airline safety. Strictly speaking, such systems go beyond the trigger event model as they pick up not only pre-defined triggers but also other, less structured information. This, however, is entirely appropriate. No organisation concerned about risk can assume that it has identified beforehand all of the potential precursors to disaster. It will therefore encourage its employees to report anything about which they feel a sense of unease.

An objection which is sometimes made to this approach is that it is only obvious with hindsight that certain events constituted warnings of what was about to happen. How can warnings be distinguished from the other signals in the environment that amount to no more than noise? The answer is implicit in the analogy. When those who are listening for signals detect what they think might be a signal, they finetune their detection apparatus to be sure of what they are hearing. Similarly, the trigger event model does not purport to identify beforehand which are the real warnings and which the false alarms. Instead, it identifies certain categories of event that call for further investigation. This may reveal that the matter is of no consequence. But the beauty of the system is that it ensures that such a judgment is made consciously and conscientiously and does not simply occur by default. Perhaps the most important point here is that signals will only be distinguished from noise if there are people whose job it is to listen and respond to what they hear. Similarly, warnings will only be realised for what they are, if there are people whose job it is to investigate them.

Finally, management must make sure that workers are provided with a variety of incentives to report. In particular, reporters must be acknowledged and thanked, and they must be provided with feedback on what is being done about their reports. Management must also be careful to ensure that there are no disincentives to reporting.

Organisational mindfulness

The TARP philosophy, and its extension into structured hazard/occurrence reporting systems, provides a resolution of sorts to one of the current safety debates. For many years, large companies have been relying on safety management systems to minimise risk. There is, however, a widespread perception that these have not worked as well as expected, that there is a disconnect between the system on paper and the system in practice, and that something more is needed. That "something more" is a safety culture (Reason, 2000). And this is where the debate comes in. On the one hand, there is a view that safety culture involves an ongoing risk-awareness or mindfulness by individuals and ongoing commitment by individuals to do things safely. Companies therefore invest heavily in programs that are aimed at developing the requisite "mindset". On the other hand, certain writers see safety culture as a truly organisational phenomenon; most importantly, a set of processes for collecting and responding to information (Reason, 1997). The TARP philosophy resolves this debate by integrating the two viewpoints. It sets up *systems* that encourage *individual* risk-awareness — systems that encourage conscientious decision-making about emerging risks. Trigger action response plans embody an organisational mindfulness that includes, but goes beyond, individual mindfulness.

This analysis suggests a way forward for companies and regulators alike. One way to encourage organisational mindfulness is to audit systems that are designed to respond to warnings of danger in order to ensure that they are working effectively. In relation to TARPs, are front-line workers aware of the trigger levels, and do supervisors really respond in the manner required by the action response plans when trigger levels are reached? Where companies have defined critical operating parameters, what really happens when the parameters are breached, and what is the company doing to reduce the occurrence of such breaches? In relation to hazard/occurrence reporting systems more generally, what kinds of things are being reported? Has management effectively alerted its workforce to what it wants reported? What is the response to reports? Do workers feel sufficiently encouraged by the response to motivate them to continue reporting? By focusing on these matters, management and regulators alike will be able to promote the kind of awareness that was lacking at Gretley.

Chapter 9

Mindful leadership

The previous chapter suggested ways in which organisations could become more mindful. The culture of an organisation depends, however, on its leaders (Schein, 1992), and a mindful organisation will require mindful leadership. The purpose of this chapter is to explore the idea of mindful leadership.

Individual mindfulness on the part of leaders goes beyond regulatory compliance. Regulations require leaders to exercise an appropriate level of care in their pursuit of safe operation. Provided they have exercised this requisite level of care, they can rest reasonably easy that they will not be personally liable if something goes wrong. Mindfulness goes beyond this. Mindful organisations, it will be remembered, exhibit chronic unease. They embrace the idea that danger may be lurking beneath the surface of normality and that, despite a record of highly reliable functioning, things could go wrong at any moment. Mindful leaders will manifest this unease. Clearly, this is not something that can be directly mandated by law.

Mindful leaders lie awake at night worrying about the possibility of a major accident. As one said to me following the Esso gas plant explosion at Longford: "How do we know we are not about to have a 'Longford' at our site? We think we are managing well, but so did Esso." Of course, worry is not always helpful and unease does not automatically generate positive outcomes. In this chapter, I want to suggest a way in which the unease felt by mindful leaders can be usefully focused. My starting point is a basic proposition from organisational sociology that "the most critical issue for organisational safety is the flow of information" (Westrum, 2004). Accordingly, the focus here will be on the flow of safety relevant information, both up and down an organisation. Mindful leaders need to be concerned about two possibilities. The first is that somewhere in the organisation information is available about trouble that is brewing — information that is not making its way upwards to people with the capacity and inclination to take effective action. In short, there is a possibility that the bad news which leaders need to know about is simply not reaching them.[1] The second possibility is about the flow of safety messages in the reverse direction. The worry is that, despite explicit statements by leaders about the importance of safety, workers are receiving a contrary message, namely, that top priority is to be given to production. This chapter will discuss how these failures in communication occur and suggest some ways of overcoming them.

---

1 This is discussed in more detail in the preceding chapter.

I shall begin by making three suggestions for improving the upwards flow of safety-relevant information. The first is that leaders must reorient their auditing procedures to probing for problems, rather than providing assurances of compliance. Moreover, leaders themselves must become personally involved in probing for problems. Second, leaders must improve their hazard and occurrence reporting systems to maximise the chance of capturing warnings of disaster. Third, they should strengthen the voices of the safety specialists in their organisations.

Probing for problems

Senior managers often see auditing as a means of providing themselves with the assurance that things are as they should be. The trouble is that, if leaders are seeking such assurances, that is what they are likely to be given. Auditors may identify what they euphemistically call "improvement opportunities", or "challenges", but if the task is to provide some overall assessment of how well the organisation is being managed, the chances are that the assessment will be positive. Even the most reputable and independent auditors will feel the pressure to provide a generally favourable audit report. They know how disheartening a negative report or a low score can be when a site has tried hard to improve its performance, and they are under considerable pressure to be supportive. The practice of providing the site with a draft report so that it can correct deficiencies before the final report goes forward is but one manifestation of this tendency to paint things in the best possible light.

Leaders who want to get beyond these appearances, and to pinpoint the unrecognised problems that may be lurking beneath the surface, need to avoid any suggestion that they are asking for assurances that their system is functioning as intended. Indeed, they need to be suspicious of audit reports that suggest that, basically, all is well. There are numerous examples of auditors reporting that all is well, only to be followed a few months later by a major accident and an inquiry that reveals that all was not well and that the auditors had failed to identify or highlight some very obvious failings (Hopkins, 2000, ch 7). Where auditors report that all is well, leaders should challenge these assurances and explore with auditors how they came to these conclusions and how much confidence can be placed in them. They might, for example, pose the following question: "Suppose there were some crucial hazards that remained uncontrolled, or some critical procedures that were not being followed. How likely is it that your audit procedures would uncover these problems?"

One way to overcome this problem is not to ask auditors to provide an overall assessment, but to ask them to identify the most significant safety issues confronting the organisation or site that is being audited. This gives a radically different purpose to auditing, and it means that auditors who fail to come up with a list of significant concerns have failed in their assignment. No longer is the end point a ranking or a set of numbers; rather, it is an agenda for action, perhaps urgent action. If everyone starts from the assumption that there are likely to be significant problems and that it is the auditor's job to identify them, then no one need feel undermined when such problems are duly identified.

This proposed audit function is so radically different from conventional auditing that it should possibly be given a different name, such as "problem detection". I shall continue to use the term "audit" in what follows, however, because the intention is that conventional auditing be transformed in this manner.

All of this can be put another way. Some leaders simply assume that their systems are working as intended. When something goes wrong, they are genuinely surprised to learn that there is a gap or a disconnect between the system in theory and the system in practice — between what people are supposed to be doing and what they are in fact doing. For instance, certain Royal Australian Air Force commanders expressed surprise when they discovered that maintenance practices at Amberley air force base near Brisbane had for years been plagued with problems of inadequate personal protective equipment and that workers had suffered serious health problems as a result. They had simply assumed that such matters would be brought to their attention. One of these commanders seemed not to recognise the problem even after it had come to light. This is what he told the Board of Inquiry:

> "I believe that the ingrained requirement to follow procedures and to supervise subordinates to ensure that this is done is such as to make instances of non-compliance the exception.
>
> I have no reason to believe that the procedures developed for the [relevant maintenance section] were not generally followed. I consider that the supervisory chain and the air force emphasis on supervision was sufficient to ensure compliance with procedures . . ."[2]

Mindful leaders, however, know that there are always gaps between the system in practice and the system in theory. The task that they give their auditors is to identify those gaps, or at least the most significant gaps, so that the organisation can work to reduce them.[3]

Auditors who are looking for problems will not approach their task as conventional auditors do. They will not set out to provide a balanced view of the organisation. Rather, they will use their expert knowledge to zero in on where things might be going wrong, and they will seek advice from others about problem areas. Moreover, they will approach their task with scepticism. They

---

2   Clarkson, J, Hopkins, A and Taylor, K, *Chemical exposure of air force maintenance workers*, Report of the board of inquiry into F111 (fuel tank) deseal/reseal and spray seal programs, Canberra, Royal Australian Air Force, June 2001, ch 3. Website at www.defence.gov.au/raaf/organisation/info_on/units/f111/Volume1.htm.

3   For example, auditing by BHP Coal following the explosion at its Moura coal mine. See Hopkins (2000:84, 85).

will not merely ask about procedures, they will observe them in action, and they will not necessarily give notice when they intend to observe what is happening, since that may alter the outcome. Indeed, auditors may need to be quite imaginative in the way that they follow up their hunches and leads to be certain that they are getting to the truth of what is going on.

From time to time, leaders may become aware of a problem at a site belonging to another company. The problem may have come to light in an accident investigation at that site. Mindful leaders will be concerned that the problem might exist in their own organisation and will appoint an individual or team to find out. For example, the explosion at the Longford gas plant near Melbourne in 1998 was initiated by a system in the control room which had so many alarms that operators could not respond effectively to them. The problem is often described as "alarm overload" or "alarm flooding". When this became known, mindful leaders in organisations around the world worried that control rooms at their sites might suffer a similar alarm overload problem, and they sent investigators to find out. They did not take comfort from the fact that routine safety assurance procedures had not identified such a problem at their site. Rather, they assumed that their routine safety assurance processes might have missed the problem and that it was therefore necessary to go looking for it specifically. Mindful leaders appoint others to be their eyes and ears and to go out and investigate — bringing to bear the same unease that they themselves feel.

Probing for problems personally

All of the procedures discussed above involve delegation, which necessarily introduces the possibility that information will be distorted or truncated. Mindful leaders know that the best way to find out what is going on is to see for themselves. This means making regular site visits to talk informally with front-line staff about the safety issues that they may be facing. But how much time should busy leaders devote to this activity? The report on the Ladbroke Grove rail crash in the UK addressed this very question:

> "Companies in the rail industry should be expected to demonstrate that they have, and implement, a system to ensure that senior management spend an adequate time devoted to safety issues, with front line workers . . . best practice suggests at least one hour per week should be formally scheduled in the diaries of senior executives for this task. Middle ranking managers should have one hour per day devoted to it, and first line managers should spend at least 30 per cent of their time in the field."[4]

Mindful leaders are aware of the importance of obtaining first-hand experience in this way. One senior coal executive I have met keeps a set of underground clothes at every mine under his control and goes underground at each of these

---

4   Cullen, L, *The Ladbroke Grove rail inquiry*, Part 2 report, Norwich, HMSO, 2001, pp 64, 65.

mines regularly. Visiting the place of work of miners underground is an enormously time-consuming business, and the time devoted by this executive to this task is an indication of the importance that he attaches to this activity.

Leaders who talk to front-line staff must find ways of engaging them and putting them at ease, but it is surprising how readily problems emerge when they do. In my own experience, asking employees on site to tell me about the problems they face elicits all sorts of safety-critical information, including examples of routine non-compliance that are necessary to get the job done.

Some companies have provided their senior executives with "prompt" questions. Here is one such set used by a major resource company:

Can you tell me about your job?

What could go wrong?

How could you prevent it?

Who else could be affected?

How can the job be done more safely?

How could you get hurt?

What kind of injury?

Such questions are likely to get conversations going. There is, however, a serious problem in the above list. The third question appears to put the onus exclusively on the employee to prevent accidents. The best way to prevent something from going wrong may be to redesign the equipment or the job, rather than assuming that the matter lies within the worker's control. Thus, the question should read: how could *we* prevent it? Such a question is far more likely to engage the employee and produce a fruitful outcome.

It is not always easy for a corporate leader to strike up a conversation with a shop-floor worker. The two may be worlds apart in social status, in ways of expressing themselves, and in knowledge and interests. These interactions are potentially awkward for both parties, and leaders may benefit from being coached or from participating in role-play, before going out into what may be for them quite a difficult environment.

Leaders who make site visits can also carry out more systematic activities. One activity is to audit some safety-critical procedure. Auditing should be done with scepticism, imagining ways in which the procedure might be failing and checking to see whether it is. The famous Piper Alpha disaster in the North Sea in 1988 began with a failure of the permit to work system. The system was regularly audited, but never sceptically, and the systematic flaws in the system were never identified. A leader who identifies flaws in a safety-critical procedure and instructs that they be corrected will have a powerful effect on the culture at that site. Where leaders are unfamiliar with procedures at the site, safety managers can assist by selecting an appropriate procedure and briefing the leader on its purpose and how it should be working.

Another systematic activity for leaders who are visiting a site is to inquire sceptically about the response to some recent accident or near-miss at the site. Examining the quality of an accident investigation and the thoroughness with which the site has implemented the relevant lessons provides an important insight into the culture at that site. Far too often, the recommended corrective action arising from an accident investigation is to *talk* to the relevant people to ensure that the problem does not occur again. This is a predictably ineffective strategy. A mindful leader will immediately recognise how such a response leaves the organisation exposed to a repeat accident.

Leaders who are doing a walk-around will often become aware of small things that are not as they should be: a missing isolation tag, a log book not properly filled out, a piece of equipment in substandard condition, or an item of personal protective equipment not being worn. The most obvious response is to ask that the matter be corrected.[5] Mindful leaders, however, are aware that these things can also be regarded as telltales (as warning signs that the management system is not functioning quite as it should be). They may therefore request an investigation into how the management system failed to pick up a particular problem and what can be done to ensure that similar problems are detected and corrected.

Finally, leaders can, in effect, use their walk-arounds to audit their own auditors. If they become aware of matters that they believe their auditors should have identified, they can feed this back in such a way as to sharpen the investigative effectiveness of their auditors.

To summarise this section, mindful leaders do not rely on assurances from subordinates that all is as it should be. They do not assume that the system is functioning as it should in theory. They know, or at least fear, that there are problems lying in wait to pounce, and they use every means available to probe for these problems and expose them before they can impact detrimentally on the organisation. They focus their audit and other investigative resources on the hunt for such problems. Moreover, they do not merely delegate this task. Mindful leaders regularly visit sites themselves and carry out their own personal audits, looking for indicators that things are not as they should be. They know that, if they delegate this investigative function entirely, they can no longer be confident that the bad news will get to them.

Improving reporting systems

The previous chapter detailed the importance of reporting systems in identifying issues and conveying information about potential problems upwards to people who can do something about them. Mindful leaders will understand that these reporting systems are essential if organisations are to pick up and respond to the

---

5 I have toured sites with corporate executives on various occasions and witnessed them responding to such matters in this way.

warnings that precede every major accident. They will be aware that such reporting systems are fragile and can easily fail to function effectively. They will therefore put considerable energy into maximising the effectiveness of such systems.

For instance, they will inquire about areas of under-reporting. Mindful leaders in the Royal Australian Navy became aware that a certain naval vessel was producing very few incident reports relative to a comparable vessel. Ironically, it is the low number of incident reports which constitutes poor performance in this context. The issue was raised with the Commanding Officer of the under-reporting ship and, within a few months, its incident reporting rate had risen.[6]

Major accidents have occurred in the past because reporting systems were narrowly focused on injury reports and ignored the kinds of events that are the precursors to disaster, such as system excursions that are beyond safe limits. Mindful leaders will therefore inspect reporting systems to see what is being reported and whether or not there are whole classes of events or conditions that are not being picked up in these systems.

The value of reports depends entirely on what is done about them. Moreover, reporters will only be motivated to continue reporting if they believe that their reports are taken seriously. Mindful leaders will thus be concerned to ensure that reports are responded to in a timely and effective manner. They may even stop production temporarily to enable the resources of the organisation to be concentrated on dealing with a backlog of matters raised in reporting systems.[7]

When leaders carrying out site visits are alerted by workers to matters of concern, they will ask whether these things have been entered into reporting systems. If not, they will want to know why not and what can be done to encourage such reporting. If the matter has made its way into a reporting system, a mindful leader will want to know why it has not yet been effectively addressed.

In short, some of the chronic unease that mindful leaders feel will be unease about how well their reporting systems are functioning. Are these systems really functioning well enough to pick up the warning signs that will enable the organisation to avoid disaster? Mindful leaders will devote part of their personal investigative activity to probing for defects in their reporting systems.

Strengthening the safety voices

Another way in which mindful leaders can increase their chance of hearing whatever bad news there may be is by strengthening the voices of those most likely to be aware of it, in particular, the safety specialists in the organisation. Large organisations have a range of people with specialist safety knowledge,

6   Clarkson et al, op cit, pp 5-7.

7   This was the practice of one leader I have spoken with.

including health and safety representatives, site safety officers, and various kinds of safety managers. However, the voices of these people can be muted by organisational structures, and they may be separated from key decision-makers by a layer of management that effectively filters or muffles their voice. At my own institution, for example, the university's organisation-wide OHS officer has usually been answerable to a human resource manager, who, in turn, is answerable to one of a group of vice-CEOs, who answers finally to the CEO.[8] In this structure, the most senior OHS specialist on campus never communicates with his CEO and almost never with anyone who even has the ear of the CEO. A structure more likely to protect the CEO from bad news on the safety front would be hard to imagine.

Mindful leaders will be aware of how reporting structures can muffle or completely silence the voice of safety specialists in their organisations. Companies that aspire to be high-reliability organisations structure themselves so that their safety staff report directly to the most senior decision-makers at every level, not via the human resources manager or some other intermediary (Parker, 2002:189). There are also reporting links between the safety staff at various levels. A simplified version of this situation is represented in Figure 1.

The implications of Figure 1 are as follows: if site A safety manager feels that supervisors are not responding to his concerns, he can speak directly to the top manager at site A.[9] If the top manager at site A does not respond, he can report this to the safety manager one level up, that is, the corporate safety manager. This is a vital part of the structure, for it means that site A manager knows that his superior (the CEO in this case) will be receiving independent reports about safety at his site. This is an excellent way of ensuring that the senior manager at each level listens attentively to any concerns that a safety manager may have. Moreover, in best practice companies, the corporate safety manager visits sites on a regular basis and, given that he is on a par with, or even outranks, the site manager, his views carry great weight. Finally, it should be noted that, if the elected health and safety representative believes that his concerns are being overridden by a supervisor, he is in a position to bypass this blockage and report directly to the site safety manager.

The distortion of the safety message

Let us consider now the problem that occurs with the downward communication of information, in particular, management's stated commitment to safety. The results of a survey in the Australian mining industry in 1999 provide evidence of the way in which leaders have failed to convey the importance that they attach to

---

8 This structure has recently changed, but the OHS manager remains separated from the CEO by two intermediate layers of management.

9 It proved impossible to write this paragraph without recourse to the male pronoun, but there may well be women in some of these positions.

FIGURE 1

Example of how safety managers can be located so as to strengthen their voice

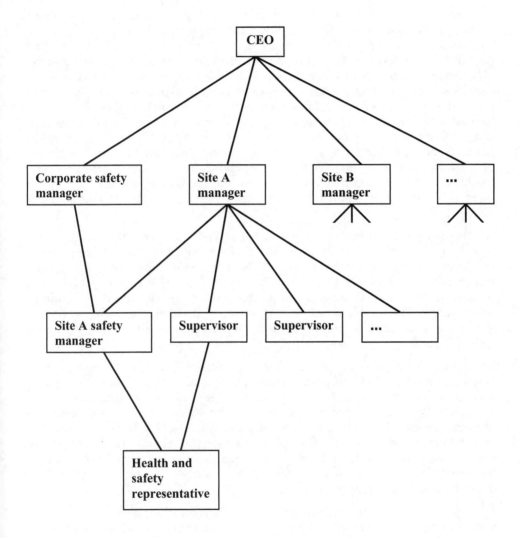

safety down the hierarchy. One survey question was: does management always put safety first? While 84% of senior managers thought that they did, only 43% of plant operators agreed. Clearly, lower-level employees did not feel that the actions of senior managers matched their words.[10]

At some sites, workers are very cynical about the commitment of senior managers. At one site where I have done fieldwork, I explored with the workforce the reasons for this cynicism. I was told that there was constant pressure to get the job done, but no corresponding pressure to do it safely. Managers routinely asked when the job would be completed and why there had been delays but, when jobs were completed in an unexpectedly short period of time, no questions were asked and people were congratulated. This is perceived as condoning any short cuts which may have been taken to get the job done. One specific example mentioned was a decision by the day shift that a cherry picker would be needed in order to safely adjust certain fittings on a tower. The cherry picker was ordered for the next day but, when the day shift arrived the next day, they found that the job had been done by the night shift without the cherry picker. The day shift concluded that their concerns about safety had been discounted.

Another job was described to me in which a procedure was regularly violated by moving a guard, and the claim made was that everyone knew about this — up to the level of the business unit leader. The perception was that management turned a blind eye to this situation. When workers have such perceptions, any statements by senior managers about the importance of safety are entirely undermined. The problem for management is that merely walking past a situation involving non-compliance will be perceived as condoning it, unless something is said.

What these examples suggest is that the messages that are inferred from behaviour are far more influential than speeches about safety; in this regard, actions speak much louder than words. This places a heavy onus of responsibility on leaders to monitor their own behaviour and to ensure that it is not inadvertently conveying the wrong message. A senior BP executive once said: "When a leader visits the workplace, they see the behaviours of their people but they also see, reflected in their people, their own behaviours."[11] Here was a leader who was acutely aware of the subtle messages that his own behaviour might inadvertently convey.

By contrast, consider the following case. The manager of a power station which was running at less than full capacity suddenly discovered that the spot price for power had risen sharply. He decided to bring two idle generators on-line

---

10   Pitzer, C, *Safety culture survey report*, Australian Minerals Industry, Canberra, Minerals Council of Australia and SAFEmap, 1999. See also Shaw and Blewitt (2000).

11   Hayward, T, BP Group Managing Director, "Working safely: a continuous journey", speech to the International Regulators' Offshore Safety Forum, London, 1 April 2005.

(without doing the normal checks), in order to take advantage of the high prices on offer. The CEO of the company which owned the power station happened to be on site that day. When the manager explained to him what he had done, the CEO thanked the manager for the concern he had shown for the company's interests but instructed him to take the generators back off-line, and do the required checks before bringing them into production again. The CEO knew that this process would take hours and would cost the company hundreds of thousands of dollars in lost revenue, but he also knew that without this intervention he would be conveying a message to the manager that, when production requirements conflicted with safety, it was acceptable to take short cuts. His action on this occasion conveyed the priority that he wanted given to safety and taught the manager an unforgettable lesson.[12]

Senior managers are not usually in direct contact with operators, so it tends to be crew leaders and shift managers who communicate to the workforce the perceived pressure to cut corners. I asked one group of these supervisors what it was about their behaviour that might convey this pressure to their crews, and the answer from one shift manager was that he "hovered" over the workers to get the job done. In so doing, he was responding to what he perceived his senior management wanted.[13] Middle managers are often blamed for distorting, even reversing, the priorities of senior managers in this way, but senior managers need to be aware of the implicit messages to which their middle-level managers are responding.

Another source of the perception that management is more concerned about production than safety is that certain jobs pose hazards that have not been designed out and that workers are expected to deal with as best they can. Sometimes these remaining hazards are potentially life-threatening. At one site I visited (operated by a reputable multinational company), management could not understand why its workforce appeared so cynical about the company's commitment to safety. It turned out that workers were exposed to some quite specific, life-threatening hazards when they operated certain furnaces. However, management had decided that the associated risks were acceptable. Such judgments are always open to debate (Hopkins, 2005, ch 12) and, in this case, workers were unhappy with management's assessment. They felt that they were in considerable danger when operating these furnaces. During my visit, I

---

12 Personal communication from the CEO.

13 I have had my own experience of just how powerful these pressures are. I was asked by a senior executive of one company to explore the safety culture at some of his mines and I was told that, if necessary, production would be stopped at the face in order to enable me to talk to miners. This is exactly what happened at one mine, and when I emerged, the mine manager greeted me with the news that the stoppage had cost them $20,000 in lost production. If that was how a visitor from corporate headquarters was treated, I wondered how the mine manager would respond to a humble miner who wanted to stop work because of safety concerns. Note that this was a mine manager who was explicitly committed to safety and who did ask me after my visit whether I had identified any safety issues while I was underground.

attended a hazard-awareness training session and, after some thoughtful discussion among participants about major risks, there was a sudden and unexpected outburst about the furnace hazards. Management's failure to eliminate these hazards was entirely undermining the credibility of its commitment to safety. In a subsequent debrief with the senior site executive, I told him that he needed to resolve this problem — even if it cost him millions of dollars. He said he agreed that the time had come to do something about it and that the hazards could probably be eliminated for a few hundred thousand dollars. Sadly, this response validated the disbelief that workers had expressed about the priority management attached to safety.

Here is a rather more mundane example of the problem. Its significance is that it comes from a coal mine where the attitude of top management is quite exemplary. At the mine in question, and no doubt at many others, in order to operate the machinery at the mine face, miners needed to move along an uneven walkway in a crouched position with heads lowered. A moment of inattention caused them to hit their heads on the machinery above. Miners told me that, no matter how careful they might be, a shift seldom went by without a crack on the head. Their hard hats protected their heads, but not their necks, which were at risk of being jarred from these blows. The perception was that nothing could be done about this situation and that this was an unavoidable by-product of the technology in use. Of course, the hazard in this case was not life-threatening. But the trade-off implicitly undermined management claims about the primacy of safety. If management is to maintain credibility, it must acknowledge to its workforce that the current situation involves a compromise, and it must seek to ensure that this hazard is eliminated from the next generation of mining machinery.

Generally speaking, if leaders are to ensure that their safety messages are not inadvertently undermined by their own behaviour, they must monitor their own behaviour carefully. If the proposition that safety is the top priority is to be effective, it must apply to the behaviour of all — from top to bottom of an organisation. Mindful leaders know that behaviour speaks louder than words, and they know that vision statements, safety slogans and admonitions are of no use unless they are matched by the behaviour of leaders themselves. Unless leadership behaviour is congruent in this way, workers end up believing that it is acceptable to take short cuts — indeed, that management expects them to take short cuts or tolerate certain hazards in order to get the job done.

Conclusion

Experience is now teaching us that safety management systems are not enough to ensure safety. What is needed, in addition, is collective mindfulness. As Weick and Sutcliffe note, collective entities that are mindful "organise themselves in such a way that they are better able to notice the unexpected in the making and halt its development" (Weick and Sutcliffe, 2001:3). As this statement makes clear, collective mindfulness is first and foremost about the style of an organisation, not about the mental state of individuals. However, it does have implications for the state of mind of individuals at all levels of an organisation. People at the front line of such organisations have an expanded awareness of risk, and leaders of such organisations display a chronic unease about the possibility of things going disastrously wrong.

This chapter has looked at what mindful leadership entails and, in particular, how mindful leaders deal with the issues of communication failure that lie at the heart of all major accidents. Mindful leaders develop procedures that maximise the chance that bad news will move up the hierarchy to people who can do something about it. They are also aware of how easily their own behaviour can convey the wrong message to their subordinates about the importance of safety.

This book has examined the role that the law can play in making leaders more safety conscious. At present, the law in various jurisdictions requires leaders to take reasonable care. I have argued that the chronic unease that mindful leaders display goes beyond the idea of reasonable care.

It is hard to see how the law could *require* leaders to behave mindfully. However, it could certainly *encourage* mindfulness. Chapter 7 argued for a regime of outcome responsibility. Under such a regime, leaders would know that they could be held personally responsible for any major accident occurring in their organisation. They would know that nothing they could do beforehand would relieve them of this responsibility. Unease about being held personally responsible would translate naturally into unease about the possibility of the accident itself. In this way, the legal system would be motivating leaders to adopt the types of strategies described in this chapter.

Chapter 10

Lessons

I have called this chapter "Lessons", not merely conclusions, because the Gretley case is so rich in implications for public policy and company practice with respect to safety. What I want to record here is not just a set of findings about what happened in the past but, more importantly, a set of recommendations for the future. These recommendations are not specific enough to be directly translated into legislation or policy. They describe the directions that law and policy need to take.

The purposes of punishment and the need for restorative justice

Policy oriented discussions about the purposes of punishment tend to focus on its crime prevention functions — deterrence, rehabilitation, incapacitation, and so on. These discussions frequently lose sight of what was perhaps the original purpose of punishment — retribution. For the retributivist, punishment is a means of ensuring that justice is done, that offenders are brought to account, and that victims or their relatives achieve some form of closure. The courts are well aware of these purposes, and they play a major part in sentencing. Though judges often make reference to deterrence, it is clear that the quantum of punishment that they impose is primarily influenced by ideas about how much punishment the offender *deserves*, not how much is necessary to ensure that deterrence is effective.

Although the public demand for prosecution in the Gretley case sometimes mentioned the need for prevention, the language was largely that of justice, accountability and closure for the relatives of those who had died. Any discussion of the merits of the Gretley prosecution must recognise and come to terms with these concerns.

Once we understand that the need for justice and closure is driving much of the public demand for punishment, we can begin to consider other ways in which these ends may be achieved — ways that may be more effective than merely inflicting pain on defendants. A criminal justice system response that focused on the needs of relatives might include face-to-face meetings between senior executives and relatives, a comprehensive apology, financial compensation, and perhaps some form of community service by relevant senior officers as penance. All this would need to be carefully negotiated among the parties. These are the kinds of outcomes suggested by the principles of restorative justice, where the aim is to restore to victims as much as possible of what they have lost, rather than simply to punish offenders. One of the strongest lessons that emerges from an examination of the demand for prosecution in the Gretley case is the need for a restorative justice approach when dealing with workplace fatalities.

Interestingly, one thing that relatives often want when loved ones are lost is an assurance that the relevant lessons have been learnt and that no one else will ever be killed in the same way. Otherwise, they say, their loved ones will have "died in vain". To know that other lives may be saved as a result of their own traumatic losses helps relatives to salvage something from the devastation. A restorative justice response is therefore not merely restorative — it can include a strong emphasis on the prevention of future fatalities. There is no inherent conflict between the preventive and restorative functions of justice.

The functions of OHS law

Occupational health and safety law aims to promote safety in the workplace. Generally speaking, it requires employers to maintain a safe workplace so far as reasonably practicable, and the real test of employer compliance is not whether an accident has occurred but whether there are hazards that the employer should have controlled but hasn't. Theoretically, prosecutions can be launched for failure to control a hazard, even though the hazard has not yet resulted in an incident of any kind. Sometimes, if the failure to control the hazard is obvious enough (as it is when there is a near-miss), the authorities may indeed prosecute. So, if a fire results from a gas escape, or if a large crane topples over, the authorities may prosecute — even though no one is harmed.[1] Normally, however, prosecutions are only launched after the hazard in question has brought about harm to workers. In this way, the function of the legislation shifts subtly from preventing harm to making employers accountable for the harm that they do to their workers.

This subtle shift in function creates enforcement problems. To convict organisations or individuals under existing OHS law requires a finding that they were at fault. In much of the legislation, being at fault means not taking the care that a reasonable person would take. That seems legitimate. But, if a manager has behaved in the way that others in his or her situation would have behaved, courts can only hold that person to account by finding that a reasonable person would have taken *more* care than is normally taken in such situations. This is exactly what happened in the Gretley case. In order to hold the individual defendants accountable for their failure to check the plans at the outset, it was assumed that the reasonable person was a person of unusual foresight who would have foreseen the possibility that the plans might be unreliable. Due diligence turned out to be exceptional diligence. This curious outcome was a direct consequence of using the law for a purpose for which it was not intended

---

1 For an interesting recent example, see *Rodney Dale Morrison v Peter Keith Ross; Rodney Dale Morrison v Glennies Creek Coal Management Pty Ltd* [2006] NSWIRComm 205.

— holding managers accountable for workplace fatalities in circumstances of minimal fault. The law was designed to hold individuals to account for failure to exercise due diligence; it was not designed to hold them to account for workplace fatalities and, in particular, it was not designed to provide closure for grieving relatives.

The elusive nature of judgments about culpability

The Gretley case also demonstrates, in a quite dramatic way, the somewhat idiosyncratic nature of legal judgments about culpability. The court found the individual defendants culpable for their failure to foresee from the outset that the plans might be in error. There are good reasons for concluding that this judgment was unfair. On the other hand, the court was not critical of the failure by managers to respond effectively in the days immediately prior to the disaster (once they had begun to suspect that something might be wrong). Using the very reasoning applied by the court, this would appear to be a more culpable failure; that is certainly the implication of the comments made by the judge who presided over the initial inquiry. Although the law purports to offer guidance for assessing the extent of culpability, it seems that culpability is not something that can be objectively determined from the facts of a situation. The divergence in the evaluations by the court and the inquiry highlights the uncertainty and, indeed, the inherent subjectivity of such judgments.

The campaign against the Act

The Gretley prosecution triggered a campaign by employers against the NSW OHS Act under which the convictions occurred. The campaign was based on three criticisms. The first was that the way the concept of reasonableness in the legislation was being interpreted was in fact unreasonable. Second, the campaign was critical of the reverse onus of proof in the NSW law; specifically, that defendants had to demonstrate that they had acted reasonably or with due diligence. The critics urged that it should be up to the prosecution to establish that they had not acted reasonably. Third, it was argued that the so-called absolute nature of the duty of care was inappropriate.

I suggested earlier that the criticism of the absolute nature of the duty of care was misconceived, and that the reverse onus of proof was not really an issue for defendants. However, there was a problem with the way in which the courts were interpreting reasonableness.

The government responded to the business campaign with a proposal to modify the wording of the OHS Act in such a way that the obligation on employers could no longer be interpreted as absolute. The proposal placed the onus of proving lack of reasonableness back on the prosecutor. None of this would have changed the outcome in the Gretley case. The government also proposed to replace the requirement that individuals act with due diligence with the requirement that they act with reasonable care. This was a semantic difference that, in and of itself, would not necessarily affect how courts interpreted the duties of individuals. The decision in the Gretley case turned on the way in which reasonableness (and due diligence) was to be understood, and the government's proposed changes did not address this issue effectively. However, there is now a discernible judicial trend towards a more commonsense understanding of the concept of reasonableness and, over time, this trend may make outcomes such as Gretley less likely. Whether the government's proposals are eventually enacted, only time will tell, but the argument here is that, by themselves, they do not preclude convictions of the Gretley type. Clearly, a more thorough response is necessary. The following sections outline such a response.

"Pure risk" prosecutions

One way that has been suggested for avoiding the problems presented by the Gretley case is to place greater reliance on what has been called "pure risk" prosecutions, that is, prosecutions in situations where an employer has failed to take action that should have been taken against some significant risk, but where no one has yet been harmed (Gunningham and Johnstone, 1999:207-209). However, these prosecutions should not be limited to dangerous occurrences that might have caused harm to people but didn't, such as the toppling crane mentioned earlier. Rather, they need to be mounted when hazardous situations that are routine in nature come to light, to continue the example, when it is discovered that the crane is routinely exceeding its safe load limits.

Prosecuting in these circumstances would ensure that the preventive functions of the legislation were not contaminated by the need to hold people accountable for death or injury. There would be little, if any, pressure from the general public to prosecute, and prosecutions would only be launched where inspectors judged offences to be flagrant. Pure risk prosecutions do not rely on the benefit of hindsight to establish that the risk is inadequately controlled. The failure must be obvious in the absence of any harmful incident. In these circumstances, only the truly negligent will be prosecuted. Moreover, there would be less need for courts to stretch the meaning of reasonableness. Finally, penalties could be tailored in a more rational way to achieve the goals of specific and general deterrence. For instance, a penalty might include a requirement that the defendant company advertise the fact of its conviction. This would heighten the shame experienced by senior executives of the company (thus enhancing the specific deterrent effect), as well as increasing employer awareness of what had happened (thus enhancing the general deterrent effect).

Making greater use of pure risk prosecutions has a number of other consequences that need to be recognised. Pure risk situations are the very situations in which inspectorates in many jurisdictions are currently using improvement notices (that is, orders that require employers to carry out certain improvements) and prohibition notices (that is, orders that prohibit an employer from carrying out, or continuing to carry out, a dangerous activity). The proposed shift requires inspectorates to escalate their response in at least some of these situations. There is good reason for this. Prohibition and improvement notices do not involve any punishment for failing to control the risk to that point. A subsequent violation of the prohibition or improvement notice is treated as an offence, but the original failure to control the risk is not. This means that there is no legal incentive for the company concerned to rectify the problem until it is drawn to its attention by the inspectorate. Furthermore, there is no legal incentive for any other company to worry about the issue until it receives a notice itself. A policy of prosecuting for pure risk means that employers have an incentive to deal with risks before the inspector calls. Pure risk prosecutions, in other words, mobilise the deterrent effect of the law in the best possible way. Moreover, because pure risk prosecutions will be imposed only in situations of considerable culpability, managers who are conscientiously seeking to control risk will not feel threatened and will be motivated to continue managing risk in a conscientious manner.

Industrial manslaughter

As we have seen, using conventional OHS law to hold people accountable for death or injury distorts the basic purpose of the law. It is here that the concept of manslaughter has a role to play. In criminal law, a person is guilty of manslaughter only if the degree of negligence is *gross*. If industrial manslaughter were to be defined in similar terms, it would make sense to refrain altogether from prosecuting individuals under OHS law in cases of workplace fatality and to prosecute instead for industrial manslaughter. It would, of course, be more difficult to establish gross negligence than mere negligence, and it is most unlikely that the Gretley defendants would have been convicted of such an offence. But where convictions *were* obtained, there would be less chance of any significant body of opinion regarding them as unfair. Moreover, there would be more chance that the truly culpable would end up in jail — as all parties seem to agree is appropriate. Of course, notwithstanding the possibility of jail sentences, there would be considerable scope for a restorative justice approach to be taken following a conviction for industrial manslaughter.

To be clear, I am not suggesting that the authorities refrain from prosecuting *companies* under OHS law when fatalities have occurred. There may be good policy reasons for doing so. The proposal is limited to individuals because it is here that the issues of fairness seem to be most acute.

It must be noted that the so-called industrial manslaughter provisions that are in place in NSW and Victoria cannot perform the function envisaged above. They require the prosecution to establish recklessness, not gross negligence. This is the degree of culpability associated with murder, not traditional manslaughter. The NSW and Victorian provisions would therefore be applicable only in exceptional cases of workplace death.

However, the preceding suggestion leaves a gap in the legal response because it will not provide the degree of individual accountability and the sense of closure that relatives need in situations such as Gretley. It is here that the issue of outcome responsibility comes into play. Indeed, it is only if the authorities are willing to impose outcome responsibility on leaders that it makes sense to refrain from prosecuting individuals under OHS law when fatalities occur.

Outcome responsibility

The principle of outcome responsibility originates in everyday life. We are responsible for the accidental harm we do, even though there was no fault on our part. If I upset you by innocently referring to some matter to which you are especially sensitive, the least I can do is apologise. There is no assumption that I am to blame, merely that I am the agent of the hurt and, in this sense, responsible. The absence of any fault on my part may mean that I do not deserve punishment, but I do have an obligation to rectify matters in whatever way I can.

This idea has its counterpart in organisational settings. There is a sense that leaders can be held responsible for what goes on within their span of control, regardless of fault. Japanese business leaders accept responsibility and resign when things go seriously wrong, even though there is no suggestion that they were directly to blame. Likewise, under an idealised system of ministerial responsibility, ministers are responsible for errors occurring in their departments and are expected to tender their resignations when those errors are serious enough.

Outcome responsibility is not as foreign in the western commercial context as it might seem. When a fatality occurs, some large, safety-conscious multinationals impose outcome responsibility on regional managers by summoning them to corporate headquarters and asking for an account of why the fatality occurred and what they intend to do to ensure that it will never happen again. What I am proposing here is that the law should be used to impose this kind of responsibility at the top of the corporate hierarchy, in particular, on CEOs.

Of course, where CEOs are held responsible for outcomes in the absence of personal fault, it would not be appropriate to impose punitive sanctions. But it would be appropriate to impose some of the consequences emphasised by restorative justice, such as face-to-face meetings between CEOs and bereaved relatives (which are aimed at satisfying some of the needs of these relatives).

Imposing outcome responsibility in this way would have a number of benefits. It would avoid the need to find individual fault, which is what made the Gretley trial so controversial. Moreover, rather than imposing responsibility on middle-level managers, it would enable the prosecution to target people at top of the corporate pyramid, that is, those who have the greatest capacity to influence corporate safety policy and practice. Corporate leaders would not be able to avoid responsibility for fatalities that occurred in their organisations by arguing that they had exercised due diligence. The court's message to them would be that, no matter how diligent they had been, they needed to try even harder to prevent such events from occurring. The general deterrent effects would be considerable, for no corporate executive would enjoy being held accountable in this way. Finally, because of the restorative justice element in this approach, imposing outcome responsibility on managers would provide some of the accountability and closure that grieving relatives want and, in so doing, would allow them to salvage a little more from the situation than they are currently able to do.

Deterrence

The preceding discussion has been primarily concerned with questions of justice and closure for relatives. I turn now to the issue of deterrence. A consequentialist justification such as deterrence is only a justification if the intended consequences actually occur. What, then, can be said about the deterrent effects of the Gretley prosecution?

When exploring this issue, we must bear in mind the distinction between the two types of deterrent effects, that is, the specific effects on the punished offender and the general effects of the wider audience on potential offenders. The judge in the Gretley case did not seek to justify punishment on the grounds of specific deterrence: the accident and its immediate consequences had had a powerful effect on the individuals concerned, and the evidence presented to the court suggested that a court-ordered punishment could not be expected to have any additional deterrent effect. Insofar as deterrence was the objective, it was general, not specific, deterrence.

General deterrence poses a moral issue that is particularly apparent in the circumstances of the Gretley case. General deterrence envisages inflicting pain of some sort on an offender — not because of any presumed effect on the offender, but to send a message to potential offenders. In short, it treats the offender as a means to an end, with all the moral problems that entails. If, in any particular case, the punishment can also be justified as necessary in order to deter the individual concerned or, alternatively, if it is justified on the grounds of desert, the moral issue is obscured. But if there is no need for specific deterrence, as was admitted in the Gretley case, and if the judgment of culpability and therefore desert is as controversial as it was in the Gretley case, the morality of using individuals as a means to an end (as general deterrence does) seems far more questionable. It is perhaps just as well that the rationale of general deterrence played a relatively small part in the determination of sentence in the Gretley case.

Nevertheless, the sentencing judge did assume that punishment would have general deterrent effects, that is, it would encourage managers, particularly in the mining industry, to attend more carefully to safety. Accordingly, I carried out a small-scale study to test this assumption. The study is described in the Appendix. It found that there were indeed deterrent effects. They were of two kinds. First, respondents reported that, as a direct result of the Gretley case, the threat of prosecution was now always in the back of their minds and that this was one factor, although often not the most important factor, that kept them focused on safety. Mine managers who were tightly integrated into large corporate structures said that corporate safety imperatives were the greatest driver of their own focus on safety. But mine managers who operated more autonomously tended to report that the fear of prosecution was a significant motivator. Second, respondents reported some practical effects — in particular, an increased tendency to write things down and an increased tendency to discipline employees for violations. While the motive for this behaviour was explicitly self-protection, the outcome was enhanced safety. Accordingly, such actions can reasonably be counted as evidence of the deterrent effect of the prosecution.

However, the general deterrent effects of the Gretley prosecution do not seem to have been as great as those reported in other studies. Part of the reason for this is that the lessons of the prosecution were not clear to other managers (apart from the need to control the risk of inrush more effectively). Managers in general were uncertain of precisely what they could do to eliminate the risk of prosecution. Where there are detailed performance standards or other prescriptive rules, managers know when they are in compliance and do no feel threatened when non-compliers are prosecuted. Indeed, they feel reassured by such prosecutions. Where managers cannot be sure that they are in compliance, the general deterrent effects on the prosecution are more muted.

A second purpose of the empirical study was to investigate claims that the Gretley prosecution had generated such a level of fear among managers and other mine office-holders that the industry was now finding it difficult to recruit people for these positions. The results indicated that this concern had been overstated in the case of mine managers and was definitely misplaced in the case of other mining positions. Where shortages did exist, interviewees attributed this to the rapid expansion of the industry rather than a fear-induced exodus from the industry.

Organisational mindfulness

The Gretley case highlights two styles of risk management. The first is to identify and control all risks at the outset. The second is the strategy of organisational mindfulness. It is based on the idea that there are always indicators of impending trouble and that mindful organisations are alert to these warnings. Indeed, mindful organisations exhibit a chronic unease about the possibility that something may be about to go wrong and they therefore put great effort into identifying the telltale signs of trouble. These two strategies are, of course, not alternatives; they are complementary, constituting a first and second line of defence against tragedy.

The prosecution viewed the Gretley disaster as essentially a failure of the first type, that is, as a failure of initial hazard identification, while the earlier inquiry emphasised that it was also a failure of the second type — a failure to attend to warning signs. As such, Gretley was a failure of organisational mindfulness.

Organisational mindfulness is a way of overcoming the limitations of safety management systems and, in particular, the disconnect between the theory of what is supposed to happen and the actual practice, because mindful organisations are naturally alert to the possibility that practices may not be as they should. A disconnect between theory and practice is a precursor to disaster, and it is precisely the kind of warning sign that mindful organisations seek to unearth.

The mining industry has institutionalised the idea of mindfulness in a particular way in its trigger action response plans but, more generally, organisational mindfulness is best institutionalised by developing systems for reporting and responding to warning signs. Reporting systems must be tailored to the particular environment and management must make clear the kinds of things that it is particularly interested in learning about. It must also find ways to encourage the kind of reporting that it wants.

Mindful leadership

Mindful organisations require mindful leadership, and mindful leaders exhibit chronic unease. Mindful leaders are concerned that there might be information available in the organisation that could help to prevent disaster, and that this information is not reaching either them or anyone else who is willing and able to take effective action. They will adopt various strategies to ferret out this information. They will use safety audits as a means of hunting for problems rather than as a means of providing assurance that all is well. Indeed, they will be sceptical of any claims that all is well. They know that bad news travels up the organisational hierarchy slowly, if at all, and they will go to the grass roots of their organisation to learn for themselves what is going on.

Mindful leaders are also acutely aware of how easily the messages that they seek to give about the priority of safety can be undermined and nullified, quite unintentionally, by their own behaviour. They know that, as far as most workers are concerned, "what interests my boss fascinates me" and that, conversely, if the boss fails to express interest in something, it is likely to receive a low priority.

It is therefore not enough for leaders to delegate safety to others; they must be seen to be actively concerned and to be behaving in ways that are consistent with that concern. Mindful leaders know that, unless they behave in this way, any statements that they make about the priority of safety will be understood as mere window-dressing.

Outcome responsibility and mindful leadership

Mindful leadership cannot be mandated by law. Chronic unease is not something that can be laid down as a legal requirement. If mindful leadership is the real key to accident-free operation, it would seem that the law has no role to play in encouraging such leadership.

This is largely true where law is fault-based. But law that is based on the principle of outcome responsibility could play a role. Recall that imposing outcome responsibility on leaders would mean that, if something went wrong, they personally would have to bear various legal consequences — no matter how careful they had been. They could not shelter behind any claim that they had exercised due diligence; they would be held accountable no matter how diligent they were. The only way that leaders could protect themselves from such legal consequences would be to ensure that nothing went wrong in the first place. Hence, unease about potential legal consequences for oneself translates naturally into unease about the possibility of something going wrong. In short, outcome responsibility generates the very unease that the leaders of mindful organisations must exhibit. There is a remarkable affinity between these two ideas. Leaders need to worry constructively about the possibility of things going wrong, and imposing outcome responsibility on them encourages them to do just that.

Appendix

Empirical study of the effects of the Gretley prosecution

Critics of the Gretley prosecution claim that it has had dire consequences for the industry. It is said that people are no longer willing to take on positions of responsibility for fear of being prosecuted, and that younger people are no longer embarking on careers in the industry because of the legal risks involved. For instance, the Mine Managers Association of Australia made the following comment in a submission to the NSW Mine Safety Review:

> "This overly punitive deterrent is causing an exodus of the more experienced and capable coal mine managers, together with other supervisory personnel from statutory positions. They are not prepared to accept the risk of prosecution when standards are impossible to meet. The qualified managers are moving to non-statutory and non-operational positions. Fewer candidates are seeking to obtain coal mining qualifications. Recent recruitment efforts by NSW coal mining companies have highlighted the acute shortage of experienced persons willing to take on supervisory and statutory roles in the industry."[1]

One observer has said, even more bluntly, that managers in the industry are in a state of "silent near-panic" as a result of the prosecutions.[2]

On the other hand, the judge who sentenced the Gretley offenders did so in the belief that the sentences would have a general deterrent effect, that is, they would focus the minds of all those in positions of responsibility in the industry on the need to be as diligent as possible in controlling risks. This, it was presumed, would have a beneficial effect on safety in the industry.

These views about the presumed effects of the prosecution are in stark contrast, but they are not inconsistent. It is conceivable that the prosecution could have had the effect of discouraging some people from accepting positions of responsibility, as well as making those who do occupy those positions more careful.

But is it the case that the prosecutions have discouraged people from accepting positions of responsibility, as industry spokespeople have claimed? And have those people in positions of responsibility become more diligent in the management of risks, as the judge intended? These are empirical questions. I

---

1   Mine Managers Association of Australia, submission to the *NSW mine safety review*, 2004, p 9.

2   "Colliery bosses cowed", *The Herald*, 16/01/06, p 13.

cannot hope to answer them definitively; to do so would be a major research project in its own right, beyond the scope of this book. But they are such important questions that they cannot be ignored. They deserve at least tentative answers here. Accordingly, I carried out a small empirical study that was designed to provide those tentative answers.

Research design

The research strategy was to interview a small sample of coal mine managers in NSW to ask them about the effects that the prosecution had had on them and on those around them. Readers with an eye to research design will recognise immediately that this design has certain limitations, for instance, it does not reach those who have left management positions and therefore it cannot yield first-hand information about their motives. Nor is it the best way to demonstrate the existence of a shortage of qualified personnel, or disentangle the factors that may have contributed to any such shortage of qualified personnel. The Mine Managers Association itself alludes to two other contributing factors in the following passage:

> ". . . the industry is enjoying a period of higher coal prices and undergoing expansion. The coal industry is rapidly moving towards a position of short supply of qualified and experienced persons willing to fill the statutory managers role . . . It is difficult enough to attract people into an industry generally regarded by the community in a negative light."[3]

This passage suggests that the expansion of the industry is pushing up demand, while a negative perception of the industry by the community is reducing supply. It may well be that these two factors together play a greater role in creating whatever shortage there may be than the fear of prosecution does. The present research design does not enable me to sort out these issues in any definitive way. Despite these limitations, the chosen research strategy can be expected to shed light on the thinking of *current* managers about the risks associated with their role. Moreover, it enables me to directly gauge the impact of the prosecution on management thinking about safety and, hence, to evaluate the general deterrent effects of the prosecution.

Coal mining is carried out in NSW in both underground and open-cut mines. Gretley was an underground mine, raising the question of whether to include both open-cut and underground mine managers in the survey. Over time, open-cut mines tend to convert to underground methods as the depth of the seam being mined increases. Furthermore, there is a great deal of communication between managers of the two types of mine and the Gretley prosecution was as much talked about by open-cut as by underground mine managers. In addition,

---

3   Mine Managers Association of Australia, op cit, p 10.

the threat of personal prosecution is not only applicable to underground mines. If the Gretley prosecution has had any of the claimed effects, we would expect to find evidence of this in open-cut as well as underground mines. For this reason, the study covers both types of mine.

The number of mines in NSW is not constant from one year to the next because some mines cease operation and others commence but, at the time of the survey, there were approximately 139 coal mines in operation. One research strategy would have been to survey managers from all mines, either by phone or by mailed questionnaire, using multiple-choice questions. This would have enabled me to draw conclusions about the percentage of managers who held particular views. However, questionnaire research of this nature elicits relatively superficial responses and I wanted to be able to explore the thinking of mine managers in more detail. Accordingly, I decided to conduct intensive, hour-long, face-to-face interviews with a small and, as far as possible, representative sample. The sample was designed to cover several geographical regions of NSW. I also wanted to include mines that were owned by large, well-known companies as well as mines that were owned by small or relatively unknown companies. Within these constraints, mines were chosen more or less at random from a list of mines contained in the NSW Coal Industry Profile.[4] Interviews were held on site, in managers' offices. Four managers whom I approached declined to be interviewed. In the end, I interviewed 13 managers in December 2005 and January 2006 (approximately nine months after the Gretley sentences were announced). Eight were from underground mines and five from open cut-mines. It should be noted that mine managers are by no means the most senior people in their companies. There was always at least one, and usually several, layers of management above them in the corporate hierarchy. Nevertheless, they are the people responsible for the day-to-day operation of the mine.

Knowledge of the prosecution

Both hypotheses about the prosecution (that it has discouraged people for taking positions of responsibility, and that it has had the deterrent effect postulated by the judge) depend on the assumption that knowledge of the case was widespread. My findings supported this assumption. All 13 managers were aware of the Gretley case. All knew that miners had broken through into old workings, that the managers had relied on plans given to them by the Department of Mineral Resources, and that these plans were in error. They also knew that two managers had been convicted and fined many thousands of dollars (although only one knew the exact figures). Just over half had believed before the sentencing that the defendants might be sent to prison. (In fact, this is

---

4   NSW Department of Mineral Resources, *2004 New South Wales coal industry profile*, St Leonards, Australia, 2004.

not a possibility for a first offence and none of the individual defendants had any previous convictions.) The Gretley case had indeed captured the attention of the industry. Moreover, open-cut mine managers were just as aware of the case — justifying the original decision to include them in the sample.

The remainder of this Appendix is divided into two parts. Part A deals with the hypothesis that the prosecution discouraged people from applying for positions of responsibility and Part B deals with the deterrence hypothesis.

Part A: the discouragement hypothesis

In view of the many public statements from the coal industry that criticised the prosecution, one might have expected mine managers to agree that the prosecution of individuals had been unfair. In fact, the great majority of interviewees (10 out of 13) took this view. However, one manager said that fatalities are always due to management system breakdowns for which someone must take responsibility, and that it was therefore appropriate that at least one Gretley manager had been prosecuted. Two other managers in my sample said that they didn't know enough to express an opinion.

Given that the two Gretley managers were generally perceived to have been unfairly singled out for prosecution, it was likely that other managers would be feeling vulnerable. After all, if the prosecuted managers were not obviously bad apples, how could the good apples be sure that they would not be similarly targeted?

In order to explore this sense of vulnerability and its impact on willingness to take on managerial responsibility, respondents were asked: did the Gretley prosecution cause you to think twice before accepting this job? All but one of the managers had accepted his current position after the prosecution of the Gretley managers had been announced, so the question was appropriate in all but one case. Somewhat surprisingly, only three managers said that they had "thought twice". One had accepted the job reluctantly and was looking for a way out. The other two deliberated carefully but concluded that the attractions outweighed the risks. Those attractions included working at a mine that was close to a metropolitan centre, the satisfaction of making career progress, and a higher salary.

Of the nine who "didn't think twice", most saw the job as a natural career progression. Certain personality attributes were also mentioned. According to one, the job required a degree of self-confidence: "You have to back your own abilities." A second said that it was "not in his nature" to worry unduly. A third described himself as not by nature risk-averse. He was, for example, a rock climber, he had learnt to hang glide, and his financial investment strategy was biased towards high-risk/high-reward investments. He can be contrasted to the manager who had been in his current position since well before the Gretley prosecution. This long-serving manager was responsible for an open-cut mine.

He had decided not to work in underground mines because of their additional hazards and he believed that working where he did reduced the risk of prosecution. His natural tendency was to avoid risks, and he explained that he invested only in blue-chip stock and budgeted conservatively. These comments raise the possibility that personality differences may influence the extent to which the prosecution impacted on managers and aspiring managers. This hypothesis could not be systematically evaluated here.

One of those who had not thought twice had changed his mind since taking the job. The mine had been poorly run prior to his appointment. On his arrival, he was confronted by a government inspector who warned him, he said, that he would be prosecuted if he made a mistake. The inspector visited the site weekly to check on compliance. The new manager succeeded in turning things around and, by the time of my interview, the inspector rarely visited the mine. But the threat lingered and the manager said that he did not expect to remain in the job indefinitely. It is clear, however, that this was a rather exceptional case. This man's fear of prosecution was a direct consequence of the explicit threats made by the inspector, rather than the Gretley prosecution itself. Of course, it was the Gretley prosecution that gave the threats their force.

This case suggests another hypothesis about why the Gretley prosecution may have discouraged some people more than others from taking on positions of responsibility. The most affected people may be those for whom the threat has been amplified in some way by relevant others. The inspector was one such relevant other who had amplified the threat. Another example of this phenomenon emerged in my study. One of the prosecuted Gretley managers had subsequently had contact with employees in positions of responsibility at a mine in my sample. The manager of this mine told me (at interview) that these employees had later expressed concern to him about the possibility of prosecution. Again, the hypothesis of threat amplification was not one that could be tested in my study.

Whether or not they had thought twice, all 13 managers concluded from the Gretley prosecution that they were at risk to some degree. However, most felt that, while the risk of being unfairly prosecuted (should there be a fatality) was high, the risk of having a fatality remained low, as long as the safety management system at the mine was functioning effectively. Nearly all drew considerable comfort from how well their systems were working. At least two managers in the sample were so confident in their systems that they viewed the risk of prosecution as insignificant.

There were several variations on this theme. One believed that his safety management system would protect him from fatalities, but that full compliance with all of the regulations could never be assured and it was inevitable that he would one day be prosecuted for some minor violation. Another said that, if he did the best he could, he could live with it, no matter what the outcome.

Two managers believed that they were fundamentally more risk-aware than the managers at Gretley and that they would not have behaved as those managers did. Thus, while not believing that the Gretley managers should have been prosecuted, they were able to distance themselves from the Gretley managers and, in so doing, to minimise their own perceived vulnerability.

We have seen, then, that three-quarters of the managers in the sample had not hesitated to apply for their current jobs. Nor were any of the 13 managers thinking of resigning because of the threat of prosecution (although two intended to move on when the opportunity arose). A substantial majority of the sample were not especially worried (indeed, were quite optimistic about their situation), believing that their safety management systems minimised the risk of prosecution.

However, as noted earlier, the research strategy was not one that would bring me face to face with people who had been dissuaded from applying for mine manager jobs because of the threat of prosecution. I attempted to deal with this problem by asking my respondents for their perceptions of how other managers or potential managers were reacting. Only one said that he knew of a manager who had retired early because of the threat. Another said that the previous manager at his mine had moved on in part, but only in part, because of the threat. A third said that he knew some engineers who did not want to be managers because of the perceived legal threat. Several interviewees said that there were relatively few qualified people available to take on mine manager positions, but that this was more a consequence of the rapid expansion of the mining industry than anything else. According to one, the industry's public position that the shortage was a result of fear of prosecution was a "furphy". Several managers noted that salaries paid to NSW coal mine managers had not risen in the way that might have been expected if the fear of prosecution was creating a serious shortage of people willing to take on the role. This point should perhaps be emphasised. Companies are perfectly at liberty to increase salaries to attract good people, just as has happened at higher corporate levels. Indeed, the salary package being paid to one of the managers in the sample was considerably above the rest, reflecting the special challenges presented by this mine.

In summary, while there are certainly some individuals who have been discouraged from applying for jobs as mine managers because of the Gretley prosecution, there was little suggestion from my respondents that this in itself had generated problems for the industry. The fact is that there are still enough qualified people to fill the available mine manager positions and the great majority of these people are enthusiastic about being mine managers.

The issue of statutory positions

Before we can investigate the discouragement hypothesis in relation to positions of responsibility below the level of mine manager, certain special features of the way in which the industry is regulated must be noted. Coal mining is subject not only to the OHS Act in NSW, but also to industry-specific legislation.[5] Under the specific legislation and its associated regulations, mine owners must create and fill certain positions, such as manager, electrical engineer, mechanical engineer and mine surveyor. These are referred to in the industry as "statutory positions". The regulations specify that these positions carry with them certain responsibilities. Many in the industry believe that occupying these statutory positions makes them particularly vulnerable in the post-Gretley environment. It is important to demonstrate that this is not the case.

The issue of whether or not the defendants occupied statutory positions was almost irrelevant in the Gretley judgment. The defendants were not tried under the industry-specific legislation but under the general OHS Act. The relevant provision of this Act at the time of the prosecution specified that individuals could only be convicted if they were "concerned in the management of the corporation". According to the judgment, the fact that a defendant was a statutory mine manager might reasonably be taken as evidence that he or she was concerned in the management of the *mine*, but not necessarily of the *corporation*.[6] That required further evidence. In the case of the Gretley managers, they were not only statutory mine managers, they were also general mine managers within the structure of the company that employed them. This implied that they had a broader range of duties that were associated with running the business, not just the functions specified under the industry-specific legislation. It was evidence about this broader role that the judge relied on, more than anything else, in finding them to be "persons concerned in the management of the corporations".[7]

Similarly, the surveyor was found to be a person concerned in the management of the corporation — not because he held the statutory position of surveyor, but because of his pivotal role in corporate decision-making. According to the judge, when a surveyor provided advice, "I have no doubt that advice would have a significant impact on decision making at the corporate level in relation to planned mining activity affecting the corporation as a whole".[8] This is what made him a "person concerned in the management of the corporation".[9]

---

5 *Coal Mines Regulation Act 1982* (NSW); *Coal Mine Health and Safety Act 2002* (NSW).

6 J 890, 886.

7 J 897, 902, 903.

8 J 941.

9 J 950. Foster (2005:123) argued that this was a mistake: "The impeccable analysis of the law concerning the meaning of the phrase 'concerned in management' offered earlier in the judgment seems to be contradicted by this finding." Foster's argument was vindicated by the appeal court. See ch 2, fn 38.

Finally, the judge determined that undermanagers (a step down from the manager in the mine hierarchy), though occupying statutory positions, were not persons concerned in the management of the corporation. Accordingly, certain undermanagers who had been charged, along with the managers, were acquitted.[10] This decision in relation to the undermanagers demonstrates conclusively that the mere fact of occupying a statutory position was irrelevant when determining liability under the OHS Act.

Notwithstanding this analysis, one of the messages that many in the industry have taken from the Gretley prosecution is that statutory office-holders are particularly at risk. The statement from the Mine Managers Association (quoted earlier) reflects this perception. The claim was that "qualified managers are moving to non-statutory and non-operational positions". The association gives no indication of how widespread this is, but it is clear that managers who make such moves will not escape liability if they remain influential in corporate decision-making.

Interestingly, most of the managers in the sample recognised that their exposure was a result of occupying managerial positions, regardless of whether they happened to be positions specified under the industry-specific legislation. The only exception was the manager who had taken on his statutory position reluctantly and was looking for a way out. In his mind, the legal risk related specifically to the statutory nature of the job. It is possible that individuals who have declined to accept positions as statutory managers have done so as a result of this misunderstanding. If that is so, the industry associations should be trying harder to educate members about the true nature of the legal risk that they face.

The impact of the prosecution on subordinate statutory positions

Given the concern in the industry about the exposure of people in statutory positions below the level of mine managers, I asked my interviewees if they were aware of any resignations at these levels. Several said that there had been talk of resignation and one manager said that he had had to persuade his electrical engineer that there was no good reason to abandon his statutory role. However, no manager was aware of a resignation that had actually occurred because of fear of prosecution.

I also asked about the difficulty of recruiting into these positions. Eight said that they had experienced no difficulty. However, one of these said that he had one qualified employee who had declined to apply for a statutory position. A ninth manager said that he had had difficulty getting certain people to apply for promotion to statutory engineering positions and that he had had to explain to them that the fact that these were statutory positions made no difference to their

---

10   J 910, 918.

legal liability. The remaining four managers said that they had experienced recruitment difficulties but that this was a consequence of an Australia-wide shortage of qualified people, especially engineers. Various observations were made about the cause of this shortage — principally, that it was a consequence of the mining boom. Several managers believed that mining companies needed to develop recruitment strategies that included training. Finally, it was suggested that mining offered a lifestyle that was increasingly unattractive compared with other available options.

In short, the situation with respect to statutory positions at levels below that of mine manager was that most mines were able to fill these positions without much difficulty, and managers who were experiencing difficulties did not attribute this to a fear of prosecution but rather to a shortage caused by industry expansion.

This last conclusion is significant. I had no direct evidence about the difficulty of recruiting people for the position of mine manager because I had interviewed mine managers themselves and not the people who appointed them. However, in the case of statutory positions below the level of mine manager, I had interviewed a representative sample of the recruiters, and they had not experienced significant difficulty in recruiting people for these subordinate statutory positions. This constitutes quite robust evidence against claims about the dire consequences of the prosecution — in particular, the claimed difficulty of recruiting to statutory positions below the level of mine manager.

Discussion

This survey suggests that the fears expressed by industry spokespeople that the prosecution had caused an exodus from statutory positions are overstated. Exodus is a strong word, suggesting that people are leaving en masse. This study has not sought to quantify the extent to which people may have made such a move, but the comments made by the interviewees do not suggest that the effect is substantial. What is clear from the survey is that there are still enough qualified and enthusiastic people to fill the available mine manager positions. It is also clear that the prosecution has not caused a recruitment problem below the level of mine manager.

This situation may change as the mining industry expands and, if the supply of qualified people available to take on positions of responsibility does prove to be insufficient, companies will be forced to increase salaries. Market forces can be expected to have this effect, regardless of the origins of the shortage. Managers therefore stand to gain financially from any shortages and, from this point of view, their association has nothing to fear.

A focus on the alleged impact of the prosecution on the availability of qualified staff detracts from what is, in some respects, a more fundamental issue, namely, the unfairness of prosecuting the Gretley defendants in the way that they were.[11] The Mine Managers Association locates this injustice in the fact that the prosecutions took place so long after the event and, more importantly, the fact that the company and its managers were prosecuted while the department that made the mistake was not. In the language of its submission to the Mine Safety Review:

> "The Company and individuals have been portrayed as the villains and the role of the Department downplayed. Where is justice? Certainly not here."[12]

However, as shown in this book, the ultimate source of unfairness is in the way in which the test of reasonable practicability has been applied. This is where the critics should be taking aim.

Part B: the deterrence hypothesis

We turn, now, to the second of the two hypotheses that this study is designed to evaluate. Before discussing the results of the present study, it is appropriate to review some of the previous research on deterrence in order to highlight the significance of the findings here.

Research generally shows that prosecuting companies for health, safety and environmental violations improves corporate performance in these areas. However, companies do not react as "amoral calculators" (Kagan and Scholz, 1984:69-72), comparing the likelihood and severity of punishment with the advantages of non-compliance. Their responses are more complex and less comprehensible from a strictly utilitarian point of view. For instance, Gray and Scholz have shown that the imposition of a penalty for an OHS offence in the United States is a shock that gains the attention of the penalised company and, to a lesser extent, other companies in the industry and that, once attention is focused in this way, companies make efforts to improve their performance (Gray and Scholz, 1991; Gray and Scholz, 1993). Furthermore, the costs of these efforts to comply may significantly outweigh the penalties for non-compliance.

Recent research on the impact of punishment for environmental offences has extended our understanding of how punishment induces compliance. Gunningham, Kagan and Thornton have carried out a series of interviews with environmental managers that explored motivations for compliance and the impact of prosecution on their thinking. This research will be outlined below because it serves as a point of departure for the present work (Gunningham et al, 2005; Thornton et al, 2005).

---

11 As noted in an earlier chapter, the prosecution might not have seemed so unfair had it been launched on the basis that the defendants failed to take effective action after they began to suspect that the plans might be in error.

12 Mine Managers Association of Australia, op cit, p 5.

Environment study by Gunningham, Kagan and Thornton

The researchers started by identifying "signal cases", that is, high-profile environmental enforcement actions, in eight different industries in the US. They then selected companies in those industries for interview.

Interviews were in two phases. In the first phase, environmental managers in 233 companies in the eight industries were surveyed by telephone, using forced-choice questions. The results are best summarised in the authors' words:

> "The survey findings suggest that most respondents did not follow closely and remember news of legal sanctions against other firms in their industry, carefully calculating their responses accordingly — as would be predicted by what we labelled 'explicit general deterrence' theory. Yet there was some support for what we labelled 'implicit general deterrence' — the sense that the mere existence of official laws and regulations entail both some risk of punishment and a duty to comply. Thus almost all respondents could remember some salient legal actions against some firms at some time in the past. And a majority reported that hearing about legal sanctions against other firms had prompted them to review, and often to take further action to strengthen, their own firm's preventive programs. For most respondents, hearing about sanctions against other firms had primarily a 'reminder' and 'reassurance' function — reminding them to review their own compliance status and reassuring them that if they invested in compliance efforts, their competitors who cheated would probably not get away with it" (Gunningham et al, 2005:290, 291).

The concept of implicit general deterrence involves an important step away from the amoral calculator assumed in some other deterrence research (Viscusi, 1979). It acknowledges that respondents feel a moral duty to comply, but it recognises that this sense of duty is contingent on the existence of a credible threat of prosecution. It is significant that regulatory research in other areas, such as taxation compliance, comes to very similar conclusions. Taxpayers are happy to comply with their obligations, provided they think others are doing the same. Penalising tax evaders provides voluntary compliers with this reassurance (Levi, 1988; Ayres and Braithwaite, 1992:26).

The second phase of the environment study involved in-depth open-ended telephone interviews with 35 company representatives in two industries — the electroplating industry and the chemical industry. The companies in the electroplating industry were nearly all small. Overwhelmingly, they believed that the threat of prosecution was a major motivator towards better environmental performance, and almost half said that they had taken action to improve environmental compliance in response to a punishment imposed on them (specific deterrence) or on some other company in the industry (general deterrence). The belief was widespread that resistance was futile, that sooner or later non-compliance would result in prosecution, and that the size of the penalties would be sufficient to threaten the viability of the business. But companies did not rationally calculate the costs and benefits of compliance; they complied with regulation because enforcement action had created a "culture of compliance, such that it becomes almost unthinkable to regulatees that they would calculatedly (as opposed to inadvertently) break the law" (Gunningham et al, 2005:309). For this reason, the authors see the deterrent effect of prosecutions in this industry as implicit rather than explicit.

As for the chemical industry, small companies exhibited the same pattern of responses as the electroplaters. In the case of larger companies, the authors found somewhat different effects. These managers were not concerned about the direct impact of fines on the business, but they were very concerned about the impact of prosecution on company reputation. A conviction and fine meant bad publicity and this could influence customers — as well as the views of the community in which the company was located. Large companies are, in effect, granted a "social licence" by their host communities (Gunningham et al, 2004) and bad publicity could turn the local community against them, jeopardising this licence. On the other hand, doing better than is required by regulation, "going beyond compliance" (Gunningham et al, 2004), makes the social licence more secure. Notice, however, that on this account the threat of prosecution is just as vital in motivating compliance in the case of large companies as it is for small companies. The difference is that, in the case of small companies, fines directly threaten financial viability, while for large companies, the imposition of penalties damages company reputation in the eyes of the local community and raises the risk of a political response that is antagonistic to the company's interests. Overall, therefore, these data provide powerful support for the deterrence thesis (especially general deterrence) in relation to corporate environmental offences.

Deterrence in the context of risk-based legislation

The environment study used a "signal case" research strategy. Gretley is clearly a signal case for the NSW coal mining industry in that it sent shock waves throughout the industry (as discussed earlier). But there are significant differences between the Gretley case and the signal cases that formed the basis of the environment study. These differences are such that it cannot be assumed that the deterrent effects seen in other studies will be found here.

In order to identify the differences, we need to first characterise in more detail the signal offences in the environment study. Seven of the eight were pollution offences in which companies discharged pollutants into the air and waterways in excess of the limits allowed by law. The eighth case involved violation of asbestos removal regulations. In all but one of the pollution cases, the discharges were systematic and occurred over prolonged periods. They were deliberate or intentional offences. The asbestos offence also appeared to be deliberate. Only one of the eight cases was an accident, in the sense of being unintentional. This was a discharge that occurred because of pipeline corrosion. This is not to say that the company was blameless: the piping had not been inspected for corrosion as it should have been.

A further aspect of these offences is that they overwhelmingly involved violations of prescriptive rules, that is, performance standards about tolerable levels of discharge.[13] Leaving aside the issues of measurement, it is, in principle, clear whether a company is in compliance with such rules: either the level of discharge is less than the allowable maximum or it is not.

Consider, now, the Gretley case. The first and obvious difference is that the offences were not deliberate or intentional. Second, the legislation was not prescriptive — it was risk-based, requiring duty holders to do what was reasonably practicable to minimise risk. This is a crucial distinction. Whether a company is in compliance is not as clear in the case of risk-based legislation (such as applied at Gretley) as it is in the case of prescriptive regulation. The question of whether a company has done all that is reasonably practicable to minimise risk is a matter of judgment and often, only after a court has ruled on the matter, is there a definitive answer. In the Gretley case, managers thought they had done all that could reasonably be expected of them to deal with the known inrush hazard (by relying on plans of the old workings supplied by the Department of Mineral Resources). It was only after the event that it became clear that there was more that they could and should have done. Jurisdictions where legislation requires companies to minimise risk may provide guidelines as to what this means in practice, with the understanding that compliance with these guidelines will be taken as evidence that risk has been reduced as far as is reasonably practicable (Bluff and Gunningham, 2004:20). But there were no such guidelines in the Gretley case.

It is worth observing that the concept of going beyond compliance (which Gunningham and colleagues discuss in the context of anti-pollution legislation (Gunningham et al, 2004)) is not as applicable where legislation is focused on reducing risk as far as reasonably practicable. No matter how praiseworthy a company's efforts to minimise risk are, the very fact that it is engaged in such efforts implies that these efforts are reasonably practicable. From this point of view, it is almost impossible to go beyond compliance. Of course, where risk-based legislation imposes performance standards or is supported by specific guidelines about what the authorities will regard as acceptable, it does become possible to go beyond compliance, just as companies can do in their response to pollution limits (Hopkins, 2007).

It is clear from this discussion that the meaning of deterrence in the two contexts under consideration is different. In the case of pollution offences, the aim is to deter people from doing what they know they should not do, namely, discharging more than the limit specified by law. In the case of risk-based

---

13   For a discussion of performance standards, see Bluff and Gunningham (2004).

legislation, there are no specific prohibitions to be observed. Rather, the legislation requires companies to take care to minimise risk. This is a far more nebulous goal. In the language of deterrence, the legislation seeks to deter carelessness or negligence, but what this means, and what companies and individuals must do to avoid a charge of negligent or insufficiently careful behaviour, is often far from clear. Whether enforcement action can effectively deter in the context of risk-based legislation is thus far more problematic than in the case of environmental prohibitions.

The issue of deterrence is further complicated by the extent to which prosecution is regarded as reasonable by relevant audiences. Where prohibitions are clear, those who choose to comply will think it reasonable to prosecute non-compliers. In these circumstances, "penalising the bad apples helps keep contingently good apples good" (Thornton et al, 2005:266). However, when it is not obvious beforehand what needs to be done to avoid prosecution, it is not possible to distinguish so clearly between good and bad apples. In these circumstances, prosecution will be perceived as unfair by all those to whom the legislation applies. This might well undermine any potential of such prosecutions to deter.

Finally, the environment study did not distinguish clearly between the deterrent effect on companies and the deterrent effect on individual decision-makers in those companies. In the eight signal cases, penalties were imposed both on companies and individuals. Moreover, interviewees spoke about the deterrent effect not only on their companies, but also on themselves (in that the fear of imprisonment was often mentioned as a motivator). This book is concerned with the issue of prosecuting individuals and so the focus here is on the deterrent effect on individual mine managers.

Specific deterrence

I shall consider, first, the deterrent effect of the Gretley prosecution on the individuals convicted. One of these individuals was the manager at the time of the accident, and the second was the manager two years earlier at the time when the departmental plans had first been accepted as accurate. The third was a surveyor who had accepted the accuracy of the plans. The sentences were handed down in 2005 (some nine years after the accident occurred), and the judgment contains information about the impact of the events at that time on these individuals.

Prior to the prosecution, all three had undergone various "rehabilitative measures". The two mine managers had completed risk management courses and the surveyor had attended various lectures and retraining seminars for surveyors. The judge's view was that all three had learnt their lesson and needed no further deterrence. The penalties that she imposed were justified on other grounds — desert and general deterrence.[14]

---

14   *McMartin v Newcastle Wallsend Coal Company Pty Ltd & Ors* [2005] NSWIRComm 31 (11 March 2005).

It is not easy to disentangle the effect of the prosecution from the effect of the accident itself. Managers who experience a fatality on their watch often report that their lives have been changed forever, regardless of any subsequent legal proceedings. The prosecution in this case certainly intensified the suffering of these three individuals, but it is hard to see how it could have had any additional deterrent effect.

There is a further consequence of these events that should be mentioned here: one manager determined never again to take on the responsibility of managing a mine.[15] His decision took effect from the time of the accident and was a consequence of his own sense of failure, rather than of the prosecution which, at that time, had not even been contemplated. Rather than motivating this individual to perform his managerial responsibilities with greater diligence, the prosecution simply confirmed his decision to abandon such responsibilities altogether. From the point of view of specific deterrence, prosecution in the case of this man was entirely counter-productive.

General deterrence

The general deterrent effects, it will be remembered, are the effects on other relevant audiences. My sample of mine managers is such an audience. In order to identify the general deterrent effects of the Gretley prosecution, it is appropriate to start with managers' views about the sources of safety.

All respondents believed that safety was a top priority at their mine and that the industry as a whole was highly safety-conscious, especially in recent years. However, they differed slightly in their explanations. Eight of the 13 mines were operated by large, well-known companies that exercised quite detailed control over mining activities. The managers of these mines all said that their own focus on safety was driven more by corporate leadership than their own legal liability. As one said, the view of corporate management that "no accident is acceptable" was in itself a powerful motivator. Another noted that the company CEO had closed a mine overseas that had had three fatalities, sending a powerful message about priorities to all mines in this multinational corporation. A third described how the CEO of his company had paid a personal visit to the mine after it had reported three high-potential events, that is, near-misses. The CEO had expressed concern to the manager that he might be "chasing coal" at the expense of safety. The manager was able to convince the CEO that the three reports were a consequence of improved reporting rather than increased risk. The manager noted, however, that the visit had made them even more safety-conscious and that the injury rate had dropped after the intervention of the CEO.

---

15   That is, a producing mine. He agreed to manage for a short time a mine that had been effectively mothballed and was not producing coal; see S 189. The other manager continued to take on management positions; see S 231, 232.

None of this is to say that the threat of personal prosecution is irrelevant. Some of the managers in this sample observed that the focus on safety (that was so evident from their corporate leadership) was, in part, a consequence of the concerns at these higher levels about personal liability. Moreover, these interviewees were well aware of their own legal exposure in the event of a serious accident. It was always in the back of their minds, they said. They all recognised that the more effectively they managed safety, the less likely they were to fall foul of the law in the way that the Gretley managers had. One of the managers in this sample described the Gretley prosecution as a message that society had sent to the mining industry about the importance of safety. Prior to Gretley, he said, the industry in NSW had exhibited a culture of risk-taking. He himself had once taken short cuts that would not be tolerated today. In this respect, he said, the Gretley prosecution had been beneficial.

Not all of the mines in the sample were operated in a hands-on way by large, safety-conscious companies. In three cases, there was very little corporate structure above the managers, and in two cases, the mine was owned by a large foreign company that had virtually no corporate presence in Australia and left its Australian mine managers to operate as they saw fit (I shall call these mines "autonomous" in what follows). The managers of these autonomous mines all said that they felt supported by their mine owner when they raised safety issues, but that there was no particular impetus to safety coming from this source. Three of these five managers said that fear of prosecution was a significant motivator. Perhaps the strongest expression of this was a manager who said: "My main aim in life is not to get prosecuted." On the other hand, a fourth manager in this group denied that the threat of prosecution had any effect on his attitude to safety. What drove his commitment, he said, was the desire to avoid a fatality, which he knew to be a personally shattering experience for a manager. The fifth manager in this group of autonomous mines told me that he did not fear prosecution because he was confident about the safety systems that he had in place. His commitment to safety stemmed, he said, from previous employment in a very safety-conscious multinational.

By way of summary, the Gretley case created a fear of being personally prosecuted and many managers reported that this fear helped to focus their minds on safety. It was not the only source of safety-consciousness, or the most important, but it was clearly influential.

Some detailed effects

The preceding conclusion is rather general in nature. In order to identify deterrent effects more precisely, managers were asked whether the Gretley prosecution had made them more likely than they might otherwise have been to take certain actions. Those actions, and the number answering affirmatively, are indicated above the dotted line in Table 1. Thus, the first row indicates that none of the 13 managers said that the Gretley prosecution had made them *more* likely to stop production for safety reasons. The action below the dotted line will be discussed shortly.

TABLE 1

Actions more likely to be taken as a result of Gretley prosecution

Stop production for safety reasons	0/13
Consult with superiors	0/13
Consult with workforce	1/13
Double-check things	3/13
Write things down	7/13
....	
Discipline violators	5/10

The figures in Table 1 suggest that the prosecution did have some specifiable effect on the behaviour of the managers — although perhaps not as much as might have been expected, given the earlier conclusion about the way in which the prosecution had focused their minds on safety. In part, the reason for this is that it was stressed to interviewees that the question was not whether they were likely to do any of these things, or whether they were more likely to do these things than previously, but whether the prosecution had made them more likely to do them. This instruction biased respondents against answering affirmatively. This would have been particularly so for managers who see their own senior executives as the primary safety driver. The influence of this primary driver could be expected to mask the influence of all other factors. So it was that all managers told me that they would stop production for safety reasons (and could give examples of doing so), but they did not attribute this behaviour to the Gretley prosecution. Three mine managers did say that they double-checked things more often as a result of the prosecution. All three were managers of autonomous mines. It is possible that, because these managers were not being driven in the direction of safety by their own superiors, they were more aware of the influence of the prosecution itself.

The strongest effect among the items above the dotted line in Table 1 is the increased tendency to write things down. Much of the motivation here was about self-protection in the event that managers find themselves in court. Written evidence that they had given certain safety instructions, had warned workers about certain things, or had "closed out" (that is, carried out) recommendations from audits and incident investigations would enable them to demonstrate that they had acted with "due diligence", as required by law. But regardless of this self-protective motivation, putting things in writing in this way makes it more likely that they are actually done. Here, then, is one very real benefit of the prosecution.

The particular actions listed above the dotted line in Table 1 were, in part, designed to get respondents thinking in these more practical terms. After probing these matters, interviewees were asked whether they could identify any other particular effects of the prosecution. One that came to light was an increased tendency to take formal disciplinary action against workers who were found to be violating safety requirements. Because this issue emerged only after interviews had begun, it was raised with only 10 respondents (see below the dotted line in Table 1). Five of the 10 reported an increased tendency to discipline workers using a formal three-stage disciplinary process, culminating in suspension or termination if necessary. Some of these managers noted that in the past, in the interests of industrial harmony, there had been a tendency not to discipline rule violators. But they now believed that failure to take action when they encountered violations by workers might be construed as condoning those violations, should the matter ever come before a court. In order to protect themselves against such an interpretation, it was necessary to take formal action, which of course was recorded in writing.

At first sight, such a response might seem almost perverse, appearing to transfer to workers the responsibility that the legislation imposes on managers. However, all of the managers who reported disciplining employees in this way also reported that they had the full support of union officials. Provided the rules were clear and were enforced in a consistent fashion, the mining union supported disciplinary action — especially when violations put the lives of others in danger. All parties recognised that the consistent enforcement of safety rules was vital if a culture of safety was to be established and maintained. A union submission to a government inquiry is quite explicit on this matter. It argues that companies that do not effectively enforce compliance with safety rules should be found guilty of negligence in the event that someone is killed as a consequence.[16] Here, then, was a somewhat unexpected outcome of the prosecution. Fear of personal liability was driving managers to respond more effectively to employee violations, with consequent benefits for safety.

A final outcome of the prosecution mentioned by four interviewees was that companies were now asking managers to involve company lawyers in the investigation of any accident. It should be noted, however, that this development is not a response by mine managers to the threat of personal liability, but a company response to the new era of prosecution ushered in by the Gretley case. There appeared to be two distinct strategies. The first was to formally place the investigation in the hands of the company's lawyers. Then, if government

---

16   Construction, Forestry, Mining and Energy Union, submission to the green paper *Transforming health and safety regulation in NSW coal mines*, 2000, p 46.

inspectors ask to see a report, lawyers can refuse to hand it over on the grounds that this violates lawyer/client confidentiality. The second strategy was to send draft reports to lawyers so that they could advise on what needed to be left out, so as to avoid self-incrimination. Interviewees complied somewhat reluctantly with these new policies because they believed that censoring reports in this way damaged relationships with local inspectors. Any such censorship of accident reports must be seen as an undesirable outcome of the Gretley prosecution which is, if anything, detrimental to safety (Hopkins, 2006).

Discussion

Previous research has demonstrated that prosecuting companies and their managers for health, safety and environmental offences has a significant deterrent effect, both specific and general. This previous research focused on deterrence in relation to prescriptive rules. These findings do not necessarily apply in the context of risk-based legislation. What needs to be done in order to comply with risk-based legislation is not as obvious and, hence, what managers need to do to avoid prosecution cannot be clearly specified. Whether prosecution in these circumstances can have a deterrent effect is thus an empirical question that is unanswered by previous research.

The Gretley case was a prosecution for failure to comply with risk-based legislative requirements. Dealing first with the issue of specific deterrence, the accident itself had a profound effect on those concerned and the prosecution had no discernable additional deterrent effect on these individuals. Indeed, the sentencing judge did not intend any such effects; the defendants had learnt the necessary lesson long ago, she said. This highlights a problematic aspect of prosecuting for failure to take sufficient care in circumstances where that failure has led to fatalities. If specific deterrence is the goal, such prosecutions are redundant. The authorities would be better off prosecuting cases where managers are failing to exercise sufficient care but no one has yet been injured. Gunningham and Johnstone (1999:207-209) call these "pure risk" prosecutions to distinguish them from cases is which the prosecution is a response not only to the ineffectively controlled risk but also to the harm done by the failure to effectively control the risk. Where managers have been careless but their carelessness has not resulted in death or injury, there will be no incentive to take greater care unless the courts impose negative consequences. These are the circumstances in which risk-based prosecutions are likely to have discernable specific deterrent effects.

Turning to the general deterrence effects of the Gretley prosecution, the present study has found that there were such effects. These were of two kinds. First, respondents reported that the threat of prosecution was always in the back of their minds and that this was one factor (although often not the most important factor) that kept them focused on safety. Second, respondents reported some very real effects — in particular, an increased tendency to write things down and an increased tendency to discipline employees for violations. While the motive for this behaviour was explicitly self-protection, the outcome was enhanced safety. Accordingly, such actions can reasonably be counted as positive outcomes of the prosecution.

However, the general deterrent effects of the Gretley prosecution do not seem to have been as great as those reported by Gunningham and colleagues. Part of the reason for this is that the lessons of the prosecution were not clear to other managers (apart from the need to control the risk of inrush more effectively). Managers in general were uncertain of precisely what they could do to eliminate the risk of prosecution. Where there are detailed performance standards or other prescriptive rules, managers know when they are in compliance and do not feel threatened when non-compliers are prosecuted. Indeed, they feel reassured by such prosecutions. Where managers cannot be sure that they are in compliance, prosecution hangs menacingly over the heads of everyone.

The pure risk prosecution advocated above is one solution to this problem. Such prosecutions do not rely on the benefit of hindsight to establish that the risk is inadequately controlled. The failure must be obvious in the absence of any harmful incident. In these circumstances, only the truly negligent will be prosecuted. Those who have taken action to deal with such obvious risks will feel assured, rather than threatened, by such prosecutions. In this way, the general deterrent effects of prosecution can be maximised.

References

Ayres, I and Braithwaite, J (1992). *Responsive regulation: transcending the deregulation debate.* Oxford: Oxford University Press.

Bagaric, M (2001). *Punishment and sentencing: a rational approach.* Sydney: Cavendish.

Bakan, J (2004). *The corporation: the pathological pursuit of profit and power.* London: Constable.

Bluff, E and Gunningham, G (2004). Principle, process, performance or what? New approaches to OHS standards setting. In Bluff, E et al (eds). *OHS regulation for a changing world of work.* Sydney: The Federation Press, pp 12–42.

Bluff, E and Johnstone, R (2005). The relationship between "reasonably practicable" and risk management regulation. *Australian Journal of Labour Law,* 18: 197–239.

Braithwaite, J (1989). *Crime, shame and reintegration.* Cambridge: Cambridge University Press.

Braithwaite, J (2002). *Restorative justice and responsive regulation.* Oxford: Oxford University Press.

Braithwaite, J and Fisse, B (1990). On the plausibility of corporate crime theory. In Laufer, W and Adler, F (eds). *Advances in criminological theory,* vol 2. Oxford: Transaction Publishers, pp 15–38.

Braithwaite, J and Pettit, P (1992). *Not just deserts: a republican theory of justice.* Oxford: Clarendon Press.

Brooks, A (1993). *Occupational health and safety law in Australia,* 4th ed. Sydney: CCH Australia Limited.

Bronitt, S and McSherry, B (2005). *Principles of criminal law,* 2nd ed. Sydney: Lawbook Co.

Brown, W (1974). Japanese management: the cultural background. In Lebra, T and Lebra, W (eds). *Japanese culture and behaviour.* Honolulu: The University Press of Hawaii, pp 174–191.

Cane, P and Gardner, J (2001). *Relating to responsibility: essays for Tony Honoré on his eightieth birthday.* Oxford: Hart Publishing.

Cressey, D (1989). The poverty of theory in corporate crime research. In Laufer, W and Adler, F (eds). *Advances in criminological theory,* vol 1. Oxford: Transaction Publishers, pp 31–55.

Dore, R (1973). *British factory–Japanese factory: the origins of national diversity in industrial relations.* Berkeley: University of California Press.

Douglas, M (1992). *Risk and blame: essays in cultural theory.* London: Routledge.

Duncan, W and Traves, S (1995). *Due diligence.* Sydney: LBC Information Services.

Duran, P (1937). *Japanese concept of official responsibility.* Manila: General Printing Press.

Epstein, R (1973). A theory of strict liability. *The Journal of Legal Studies*, 2: 151–204.

Finer, S (1956). The individual responsibility of ministers. *Public Administration*, 34: 377–395.

Fisse, B (1990). *Howard's criminal law*, 5th ed. Sydney: Lawbook Co.

Fisse, B and Braithwaite, J (1993). *Corporations, crime and accountability*. Cambridge: Cambridge University Press.

Flyvberg, B (2001). *Making social science matter*. Cambridge: Cambridge University Press.

Foster, N (2005). Personal liability of company officers for corporate OHS breaches: section 26 of the OHS Act 2000 (NSW). *Australian Journal of Labour Law*, 18: 107–135.

Foster, N (2006). Manslaughter by managers: the personal liability of company officers for death flowing from company workplace safety breach. *The Flinders Journal of Law Reform*, 9: 79–111.

Flemming, J (1984). Is there a future for tort? *The Australian Law Journal*, 58 (March): 131–142.

Friedman, W (1972). *Law in a changing society*. Harmondsworth, England: Penguin Books.

Gardner, J (2001). Obligations and outcomes in the law of torts. In Cane, P and Gardner, J (eds). *Relating to responsibility: essays for Tony Honoré on his eightieth birthday*. Oxford: Hart Publishing, 111–143.

Glasbeek, H (2005). More criminalisation in Canada: more of the same? *The Flinders Journal of Law Reform*, 8(1): 39–56.

Gobert, J (2005). The politics of corporate manslaughter — the British experience. *The Flinders Journal of Law Reform*, 8(1): 1–38.

Gray, W and Scholz, J (1991). Analysing the equity and efficiency of OSHA enforcement. *Law & Policy*, 13(3): 185–214.

Gray, W and Scholz, J (1993). Does regulatory enforcement work? A panel analysis of OSHA enforcement. *Law and Society Review*, 27(1): 177–213.

Gunningham, N and Johnstone, R (1999). *Regulating workplace safety: systems and sanctions*. Oxford: Oxford University Press.

Gunningham, N, Kagan, R and Thornton, D (2004). Social license and environmental protection: why businesses go beyond compliance. *Law and Social Inquiry*, 29(2): 308–341.

Gunningham, N, Thornton, D and Kagan, R (2005). Motivating management: corporate compliance in environmental protection. *Law & Policy*, 27(2): 289–316.

Hall, G (1994). *Sentencing guide*. Wellington, NZ: Butterworths.

Hart, H and Honoré, T (1985). *Causation in law*, 2nd ed. Oxford: Clarendon Press.

Honderich, T (1971). *Punishment: the supposed justifications*. Harmondsworth, England: Penguin Books.

Honoré, T (1999). *Responsibility and fault.* Oxford: Hart Publishing.

Hopkins, A (1999). *Managing major hazards: the lessons of the Moura mine disaster.* Sydney: Allen & Unwin.

Hopkins, A (2000). *Lessons from Longford: the Esso gas plant explosion.* Sydney: CCH Australia Limited.

Hopkins, A (2002). *Lessons from Longford: the trial.* Sydney: CCH Australia Limited.

Hopkins, A (2002a). Two models of major hazard regulation: recent Australian experience. In Kirwin, B, Hale, A and Hopkins, A (eds). *Changing regulation: controlling risks in society.* London: Pergamon.

Hopkins, A (2005). *Safety, culture and risk.* Sydney: CCH Australia Limited.

Hopkins, A (2006). A corporate dilemma: to be a learning organisation or to minimise liability. *The Journal of Occupational Health and Safety — Australia and New Zealand,* 22(3): 251–259.

Hopkins, A (2007). Beyond compliance monitoring: new strategies for safety regulators. *Law & Policy,* 29(2): 210–225.

Janis, I (1972). *Victims of groupthink: a psychological study of foreign-policy decisions and fiascoes.* Boston: Houghton Mifflin.

Johnstone, R (2004). *Occupational health and safety law and policy,* 2nd ed. Sydney: LBC Information Services.

Kagan, R and Scholz, J (1984). The "criminology of the corporation" and regulatory enforcement strategies. In Hawkins, K and Thomas, J (eds). *Enforcing regulation.* Boston: Kluwer-Nijhoff.

Kam, C (2000). Not just parliamentary "cowboys and Indians": ministerial responsibility and bureaucratic drift. *Governance,* 13(3): 365–391.

Levi, M (1988). *Of rule and revenue.* Berkeley: University of California Press.

Luntz, H and Hambly, D (2002). *Torts: cases and commentary,* 5th ed. Sydney: LexisNexis Butterworths.

Parker, C (2002). *The open corporation.* Melbourne: Cambridge University Press.

Parry, D (1995). Judicial approaches to due diligence. *Criminal Law Review,* Sept: 695–701.

Reason, J (1997). *Managing the risks of organisational accidents.* Aldershot, UK: Ashgate.

Reason, J (2000). Beyond the limitations of safety systems. *Australian Safety News,* April: 54–55.

Schein, E (1992). *Organisational culture and leadership,* 2nd ed. San Francisco: Jossey-Bass.

Shaw, A and Blewitt, V (2000). What works? The strategies which help to integrate OHS management within business development and the role of the outsider. In Frick, K et al (eds). *Systematic occupational health and safety management.* Amsterdam: Pergamon, ch 21.

Sheley, J (2000). *Criminology: a contemporary handbook*. Belmont, CA: Wadsworth.

Simons, K (1997). When is strict criminal liability just? *The Journal of Criminal Law and Criminology*, 87(4): 1075–1137.

Snook, S (2000). *Friendly fire*. Princeton, New Jersey: Princeton University Press.

Spigelman, J (2002). Negligence: the last outpost of the welfare state. *The Australian Law Journal*, 76 (July): 432–451.

Spigelman, J (2003). Negligence and insurance premiums: recent changes in Australian law. *Torts Law Journal*, 11(3): 291–311.

Stephan, S (2001). Decision-making in incident control teams. *The Journal of Occupational Health and Safety — Australia and New Zealand*, 17(2): 135–145.

Thornton, D, Gunningham, N and Kagan, R (2005). General deterrence and corporate environmental behaviour. *Law & Policy*, 27(2): 262–288.

Thompson, W (2001). *Understanding NSW occupational health and safety legislation*. Sydney: CCH Australia Limited.

Turner, B (1978). *Man-made disasters*. London: Wykeham.

Viscusi, W (1979). The impact of occupational safety and health regulation. *The Bell Journal of Economics*, 10: 117.

Weick, K and Sutcliffe, K (2001). *Managing the unexpected: assuring high performance in an age of complexity*. San Francisco: Jossey-Bass.

Weick, K, Sutcliffe, K and Obstfeld, D (1999). Organising for high reliability: processes of collective mindfulness. *Research in Organisational Behaviour*, 21: 81–123.

Wells, C (2001). *Corporations and criminal responsibility*, 2nd ed. Oxford: Clarendon Press.

Westrum, R (2004). A typology of organisational cultures. *Quality and Safety in Health Care*, 3(suppl II): ii22–ii27.

Index